BODY CULTURES

Body Cultures explores the relationship between the body, sport and landscape. This book presents the first critically edited collection of Henning Eichberg's provocative essays into 'body culture', enquiring into the themes of space and place through considerations of the spatial dimensions of the body, culture and sport in society. Eichberg, a well-known scholar in much of continental Europe who draws upon the diverse ideas of Elias, Foucault and others, is now attracting considerable interest from Anglo-American scholars in the humanities and social sciences.

Body Cultures is a unique collection of Eichberg's most significant writings, extensively edited to highlight his most important arguments and themes. The editors focus particularly on Eichberg's challenging claims about the notion of space: from the micro-scale of how human bodies 'express' themselves or are formally 'disciplined' through their movements in space, to the macro-scale of how bodies and cultures are invented and contested in connection with the self-identities which they come to possess in given places, regions, territories and nation-states. Introductory essays from the editors and Susan Brownell provide clear explanations and interpretations of key themes, as well as an interpretative biography of Eichberg.

Body Cultures presents the first systematic 'reading' of Eichberg's work to be published in English, enabling readers to access and interpret his innovative ideas on 'body-cultures' for the first time, and suggesting fresh ways to conceptualise the transitions from pre-modernity to modernity and post-modernity.

John Bale is Reader in Geography and Education at Keele University and **Chris Philo** is Professor of Geography at Glasgow University.

BODY CULTURES

Essays on sport, space and identity

Henning Eichberg

Edited by John Bale and Chris Philo

With a contribution by Susan Brownell

London and New York

First published 1998
by Routledge
2 Park Square, Milton Park, Abingdon, Oxon, OX14 4RN

Simultaneously published in the USA and Canada
by Routledge
270 Madison Ave, New York NY 10016

Reprinted 2000 (twice)

Transferred to Digital Printing 2006

Routledge is an imprint of Taylor & Francis Group

Typeset in Garamond by
RefineCatch Limited, Bungay, Suffolk

British Library Cataloguing in Publication Data
A catalogue record for this book is available from the British Library

Library of Congress Cataloging in Publication Data
Eichberg, Henning.
[Selections. English. 1997]
Body Cultures: essays on sport, space, and identity by Henning Eichberg/edited by
John Bale and Chris Philo with a contribution from Susan Brownell
p.   cm.
Includes bibliographical references and index.
1. Sports—Sociological aspects.   2. Identity (Psychology)
3. Body, Human.   4. Physical education and training—Social aspects.
5. Postmodernism.   I. Bale, John.   II. Philo, Chris.
III. Brownell, Susan.   IV. Title.
GV706.5.E54213   1997
306.4′83–dc21   97–19206
ISBN 0–415–17232–2

Publisher's Note
The publisher has gone to great lengths to ensure the quality of this reprint
but points out that some imperfections in the original may be apparent

CONTENTS

List of figures   vii
Notes on contributors   ix
Acknowledgements   xi

Introductory essays

1 INTRODUCTION: HENNING EICHBERG, SPACE,
   IDENTITY AND BODY CULTURE   3
   John Bale and Chris Philo

2 THINKING DANGEROUSLY: THE PERSON AND
   HIS IDEAS   22
   Susan Brownell

Essays by Henning Eichberg

The body in space

3 THE ENCLOSURE OF THE BODY: THE
   HISTORICAL RELATIVITY OF 'HEALTH',
   'NATURE' AND THE ENVIRONMENT OF SPORT   47

4 NEW SPATIAL CONFIGURATIONS OF SPORT?
   EXPERIENCES FROM DANISH ALTERNATIVE
   PLANNING   68

Bodies, cultures and identities

5 SPORT IN LIBYA: PHYSICAL CULTURE AS AN
   INDICATOR OF SOCIETAL CONTRADICTIONS   87
   with Ali Yehia El Mansouri

6 OLYMPIC SPORT: NEO-COLONIALISM AND
   ALTERNATIVES   100

CONTENTS

Towards a new paradigm

7 BODY CULTURE AS PARADIGM: THE DANISH
   SOCIOLOGY OF SPORT   111

8 A REVOLUTION OF BODY CULTURE?
   TRADITIONAL GAMES ON THE WAY FROM
   MODERNISATION TO 'POSTMODERNITY'   128

9 THE SOCIETAL CONSTRUCTION OF TIME AND
   SPACE AS SOCIOLOGY'S WAY HOME TO
   PHILOSOPHY: SPORT AS PARADIGM   149

   Index   165

FIGURES

5.1 Libya: sport and society in comparison   96
7.1 A 'trialectic' of sports   124
8.1 Premodern, modern and postmodern forms of games, sports and
    body cultures   145
9.1 Sack racing   150
9.2 Track racing   150

NOTES ON CONTRIBUTORS

John Bale is Reader in the Department of Education at Keele University. In 1994 he was visiting professor at the University of Jyväskylä in Finland and has lectured on social, historical and geographical aspects of sports in many universities in Europe and North America. He has authored (among many books and articles) *Sport, Space and the City* (London, 1993), *Landscapes of Modern Sport* (London, 1994) and (with Joe Sang) *Kenyan Running: Movement Culture, Geography and Global Change* (London, 1996). His current research is focused on the representation, in written texts and photographs, of early twentieth-century African corporeality and athleticism.

Susan Brownell was a nationally ranked athlete in the United States (in the heptathlon) before winning a gold medal for Beijing City in the 1986 National College Games during a year of language study at Beijing University. She also studied sport theory at the Beijing University of Physical Education (1987–8). She is Assistant Professor of Anthropology at the University of Missouri, St Louis, and is author of *Training the Body for China: Sports in the Moral Order of the People's Republic* (Chicago, Ill., 1995).

Henning Eichberg is a cultural sociologist and a research fellow at Idrætsforsk, the Research Institute for Sport, Body and Culture in Gerlev, Denmark. He received his PhD in history in 1970 at Ruhr University, Bochum and his habilitation degree in cultural sociology in 1976 at Stuttgart University. In 1982 he emigrated to Denmark, where he has held professorships at the universities of Odense and Copenhagen. He has also been a visiting professor at the universities of Vechta/Osnabrück, Berlin, Jyväskylä, Salzburg, Rennes and Graz, and was founder of the Institut International d'Anthropologie Corporelle. He has authored and co-authored 30 books in the fields of the history, sociology and psychology of body culture and sport, the history of military technology, Indonesian studies and studies in ethnic minorities and national identity.

Chris Philo is presently Professor of Geography in the Department of

Geography and Topographic Science at the University of Glasgow. He has co-authored *Approaching Human Geography: An Introduction to Contemporary Theoretical Debates* (with David Sadler and Paul Cloke, London, 1991), edited *Off the Map: The Social Geography of Poverty in the UK* (London, 1995), compiled *New Words, New Worlds: Reconceptualising Social and Cultural Geography* (Lampeter, 1991) and co-edited *Selling Places: The City as Cultural Capital, Past and Present* (Oxford, 1993). His specialist research is on the historical geography of 'madness' and 'asylums', taking seriously the socio-spatial construction of mental ill-health and its treatment settings.

ACKNOWLEDGEMENTS

It is difficult to know how to thank Henning Eichberg for his assistance in the completion of this project: he is, of course, the inspiration behind the whole thing, but has also been involved throughout in checking our edited versions of some of his writings and in trying to ensure that this collection of his essays in English is as well put together as possible. We are also extremely grateful to Susan Brownell for undertaking the difficult task of writing the biographical essay on Eichberg that follows our introductory chapter. In addition we must thank Nigel Thrift for his support when it was most needed.

We acknowledge permission from the respective publishers to reprint, in edited form, the essays by Henning Eichberg that first appeared in the following publications: chapter 3 in *Journal of Contemporary History*, Vol. 21, 1986, pp. 99–121; chapter 4 in *International Review for the Sociology of Sport*, Vol. 28, 1993, pp. 245–63; chapter 5 in H. Ueberhorst (ed.), *Geschichte de Leibes-übungen* (Berlin, 1989), pp. 261–73; chapter 6 in *International Review for the Sociology of Sport*, Vol. 19, 1984, pp. 97–104; chapter 7 in *International Review for the Sociology of Sport*, Vol. 24, 1989, pp. 43–60; chapter 8 in J.-J. Barreau and G. Jaouen (eds) *Éclipse et Renaissance des Jeux Populaires* (Rennes, 1991), pp. 101–29; chapter 9 in K. H. Bette and A. Rütten (eds) *International Sociology of Sport. Festschrift in Honour of Günther Lüschen* (Stuttgart, 1995), pp. 111–29.

Thanks are also due to Steven McGinley and Mike Shand at the Department of Geography and Topographic Science, University of Glasgow, and to Andrew Lawrence at the Department of Geography, Keele University, for help in the scanning and preparation of diagrams; and to Oliver Valins for compiling the index.

John Bale and Chris Philo
Keele and Glasgow
February 1997

INTRODUCTORY ESSAYS

1

INTRODUCTION
Henning Eichberg, space, identity and body culture
John Bale and Chris Philo

HENNING EICHBERG: WORK AND RECEPTION

Henning Eichberg's publications are many and varied: 'brilliant and prolific', according to the influential American sports historian, Allen Guttmann (1988, p. 209). His longest (and some would say his most significant) works have been published in German and Danish (Eichberg 1973, 1978, 1988; Eichberg and Jespersen 1985), and it is mainly in the form of articles that his work has appeared in English. Some of these papers are readily accessible; others are scattered in journals and somewhat obscure collections and conference proceedings across a very wide disciplinary spectrum. It is the breadth and hence the relative inaccessibility of this corpus of knowledge, relevant to students in a variety of disciplines in the humanities and social sciences, that we feel provides the *raison d'être* for the present collection.

Eichberg's early work was based on his training as an historian (see Susan Brownell's essay, which follows). American scholars, notably Guttmann and Richard Mandell (1984), clearly acknowledge their debt to him in their overviews of the history of sport and the essential differences that they recognise between modern sport and its folk-game antecedents. In the last two decades Eichberg's disciplinary background has become increasingly difficult to identify, however, and he has worked vigorously with concepts that would be recognisable to historians, anthropologists, sociologists, geographers, philosophers, architects and educationists. His recent and current work is, therefore, most satisfactorily accommodated under the umbrella term of 'cultural studies'.

Although Eichberg would see his work as having broad relevance to the very essence of modern and late-modern society, he seems to have been labelled principally as a major figure in the academic study of sport. This he certainly is. But to restrict his influence – or, at least, his potential influence – solely to sports studies would be misleading. For example, his scholarly work has extended beyond sports to include work on several other substantive

3

themes. His early scholarly work focused on the history of the relationship between seventeenth-century military organisation and war technology, including fortifications (Eichberg, 1976). He has also written on the distinctive characteristics of the Danish educational system, in particular the tradition of the *folkehøjskole* and the founder of the folk high school movement, N. F. S. Grundtvig (Eichberg, 1992). Indeed, in several of the essays that follow, Grundtvig (1783–1872) can be seen as an important influence on Eichberg's own work and philosophy. In addition, Eichberg's interests have extended into the fields of 'performance' and 'spectacle', as reflected in his work on the mass theatrical stagings by Nazis and left-wing movements in 1930s Germany (Eichberg *et al.*, 1977). It would hence be a crude oversimplification to assume a narrow interest in the social and humanistic study of sports, even if it is in these fields that Eichberg's work is most well known, and even if it is these subjects which – on the face of it – form the foci of the chapters that follow.

Another reason for caution in identifying him solely as a student of sports is that it could be reasonably claimed that the starting point in his many studies of 'body culture' or 'movement culture' lie not in a taken-for-granted assumption about what 'sport' is, but in a recognition of very many different configurations of the human body. The 'sportised' body may assume several such configurations. 'Serious sport' (or 'elite sport' or 'achievement sport') is only one of several possible configurations in modernity. Eichberg applies his notion of a 'trialectic' in his desire to avoid the use of simple dualisms (e.g. sport/leisure) and to avoid a vulgar interpretation of 'sport'. This 'trialectic', introduced implicitly in the first of his essays presented below, amounts to an 'ideal type' for providing new and critical insights on body culture. This idea, which is present in most of his work since the early 1980s, is presented more formally in Chapter 7.

Eichberg's work has not been entirely ignored by scholars outside the multidisciplinary study of sports, as exemplified by his dazzling contribution to a substantive collection of essays on *Fin de Siècle and its Legacy* (Eichberg, 1990a), but it is undoubtedly in relation to research on the history and geography of sports that his ideas have been most obviously adopted. His influence is also present in sociological studies of sports, although oddly he does not appear as a major figure in recent debates involving the 'figurationist' followers of the work of Norbert Elias. Hence, his work is barely mentioned in two recent collections edited by the 'Leicester School' (Dunning and Rojek, 1992; Dunning, Maguire and Pearton, 1993), nor in discussions surrounding the relevance to sports of the Giddens-inspired theme of structure and agency (Gruneau, 1993). There is certainly a paradox here. Eichberg's writings draw not only on Eliasian thinking, it is clearly 'configurational', making interconnections at a variety of geographical scales and across numerous historical contexts. This sometimes involves the execution of highly imaginative contextual and conceptual leaps, many examples of which are found in the pages that follow.[1] In the extensive review of sociological studies of sport covered in *Sport and Leisure in Social Thought* (Jarvie and Maguire, 1994), Eichberg's work is only given a passing mention. Yet these omissions do not seem to square with his willingness to seek more humanistic forms of discourse in the social sciences of sports, as reflected in his editorship in 1994 of a special issue of the *International Review for the Sociology of Sport* on the theme of a 'narrative sociology of sport'.

In our collection we have chosen to focus on aspects of Eichberg's post-1980 work that relate to space and place. This tactic is largely the result of our geographical background and our special concern for questions about sport and space and the social disciplining of bodies in space. This is why geography is such a reference point in this chapter, but it also signals to those in cognate disciplines the presence of a varied geographical literature that may be relevant to their work. From our own perspectives, Eichberg can be said to provide fresh insights into spatial and environmental aspects of society, as exemplified in the context of modern and premodern forms of body culture. First, however, we feel that it is worth briefly reviewing the contributions of geography to the study of sport, and second, to make some observations about approaches to the inclusion of space in the study of the body.

SPACE, PLACE AND SPORT: BRIEF THOUGHTS

Traditional studies of both the history and sociology of sport have tended to view the world as one-dimensional. Indeed, as recently as 1991, it was noted in a review of Eichberg's book *Leistungsräume* that the 'spatial environment' in sports research is a 'really alarming deficiency' (Digel, 1991). Although this understates the amount of 'geographical' work that has been authored on sports, both by geographers *qua* geographers and by scholars in cognate disciplines, it remains true that there *is* a deficiency of a particular kind of 'geographical' work applied to sport.

In the last decade or more, geographical studies of sport have moved away from an emphasis on choropleth mapping, which characterised most trad-itional approaches, and have begun to lean more towards both welfare and humanistic perspectives. For example, a large number of studies have been undertaken on the spatial and environmental impacts – positive and negative – of sports events on communities that surround the stadiums and arenas where they take place (see Bale, 1993). These approaches overlap with work by economists and their applications of regional multiplier models to sports stadium construction and sports franchise relocation (e.g. Baade, 1995). A more humanistic skein is drawn by geographical approaches to the sports landscape (Bale, 1994; Raitz, 1995). Some of this work, it is true, draws inspiration from the essays authored by Eichberg that are found in the early parts of the present collection, but much is traditional in approach, lacking both theoretical underpinning and critical analysis. Despite these trends, it is probably true that most sports–geographic writing – certainly in the United States – remains rooted in what might be termed a 'cartographic fetish' (Bale, 1992). By this we mean the 'scientific recording' (mapping) or modelling of geographical patterns of sports participation and the resulting recognition of 'sports regions'. We do not wish to deny the worth of such work, best exemplified in a superb and diligently produced *Atlas of American Sport* (Rooney and Pillsbury, 1992). Far from it, since such meticulously produced studies are not only intrinsically interesting in highlighting the geographical mosaic of the world of sport; they are also valuable for purposes of planning and policy. Yet such approaches are not without problems.

As we see it these problems are threefold. First, they tend to reduce people to dots or flow lines on maps; human beings become passive ingredients of, for example, a gravity model. The 'meaning' of sport in place is neglected and oversimplified. The maps set up the questions but do not begin to look for answers. A second problem with traditional studies in the geography of sport is that they tend to fragment the subject of study, which serves to isolate it from broader themes and influences. Hence, rather than work within the broad field of cultural geography, sports–geographic studies become isolated within a specialist sub-discipline with its own journal and 'speciality group'. There is, however, an interesting dilemma here. By studying sport, for example, do geographers (for example) seek to expand our knowledge of sports or of the society – or the geography – within which sports are embedded?

The geographical literature reveals, perhaps, only one paper and one part of a major book that adopts the latter approach. Both happen to be authored by Allan Pred. The former is his study of baseball fandom's 'journey to spectate' in late nineteenth- and early twentieth-century America. In this study Pred (1981) identifies the changing time-constraints on workers' spatial (and recreational) behaviour in the burgeoning industrial metropolis – *in the context of baseball*. The second example, also by Pred (1995), is his work on the Stockholm Globe, around which is constructed one chapter in his *Recognizing European Modernities*. The Globe, a major sports and recreational facility in Stockholm, is seen by Pred as one of several landscapes of spectacle that the city has housed over the past century. Pred uses the Globe not only to reveal metaphorically the way in which the global phenomenon of modern sport impacts, through the construction of such a site of spectacle, on the social and political life of the city, but also to demonstrate its significance as a site for the adoration of 'embodied commodities'. Such approaches not only inform our knowledge of the nineteenth- and late twentieth-century cities respectively, and in so doing illustrate the salience of using 'time–geographic' concepts to probe the constitution of (post) modernity; they are also valuable in opening the eyes of scholars to the viability of various facets of sports as appropriate research themes.

The paradoxical result of the myopic nature of much sports–geographic

6

writing is that what are arguably the 'best' geographies of sports are often written by 'non-geographers'. A recent example is that of the cultural anthropologist Charles Springwood's (1996) brilliant study of the iconography of two baseball places in the United States, Cooperstown and Dyersville. This not only draws on the work of the geographers David Harvey (1989) and Edward Soja (1989), but is also able to construct an account of different American pastoral dreams by weaving their ideas around studies from the broader terrains of social and cultural theory. Other examples are provided by Eichberg's writings, so we believe, and in the pages that follow, notions of landscape, place, ecology and globalisation are all approached within the context of Eichberg's distinctive take on body cultural analysis.

Recent developments in human geography have been more closely allied to social, cultural and literary theory than to the study and mapping of 'material culture', and a third problem with the traditional approach to spatial studies of sports is arguably its failure to draw on the intellectual gains of (what has been termed) this 'new cultural geography'. To take one simple example, the tremendous impact that post-colonial studies have had on cultural geography is hardly reflected in the geographical study of sports, clearly one of the prime legacies of colonialism and imperialism (Bale, 1996; Bale and Sang, 1996). The agenda for the new cultural geography would seem to possess the potential for at least exploring the world of sport, not only to inform our thinking on it but arguably also to regard it as a 'paradigm' for our times. Consider, for example, the view of Denis Cosgrove and Ali Rogers (1991), who believe that global culture is one theme (there are many others) that 'could serve as the object of a broadened social and cultural geography'. It could be argued that among the most visible forms of global culture today is that of sport, and J. Galtung (1984) has treated sport as an 'isomorph' of the world system. Under the broad rubric of 'global culture', modern sport illustrates several of the prescriptions for a globally sensitive cultural geography as listed by Cosgrove and Rogers. One is the notion of 'westernisation' (see Eichberg's essay on 'Olympic sport': Chapter 6 below); another is the proliferation of 'global cultural experiences, expressions and events', among which they explictly include the Olympics, the World Cup and Sport Aid. But, as Eichberg points out in Chapter 8, this proliferation also includes a whole host of 'alternative' global, regional and local 'events' and gatherings that not only mimic, but also seem on occasion to react against, Olympic monumentalism.

Sport is central to a number of other concepts cited as focal points in an agenda for cultural geographic studies. Cosgrove and Rogers also see the area of 'territoriality and nationalism' as a theme that a new cultural geography might address, for instance, and Eichberg's insights into the many faces of territorialised and segmented space (see Chapters 3 and 4) are valuable in this respect. Sport is a world of territoriality, while representational sport draws on and amplifies nationalist feeling. In international sports events, national symbolism is 'over explicit' (Ehn, 1989). Among the sub-themes identified by Cosgove and Rogers are 'myths of nation', and in constructing 'myths of nation' sports may form a central role (and Cosgrove and Rogers actually allude to Norman Tebbit's 'cricket test', which may not be as 'petty' as they suggest). An awareness of Eichberg's ideas about different configurations of 'body culture' may be central in recognising national assertiveness in multicultural societies. Various non-sportised forms of 'movement culture' may also become central in national assertiveness in 'supra-national times'. It is often through dance or sport that the identities of minority groups or the populations of newly emergent states are consolidated (see several of Eichberg's contributions to this book). Seeking the ways in which bodies – and landscapes – are configured is a means of recognising diversity in what is often thought to be an increasingly homogeneous world.

SPACE, HISTORY AND BODY CULTURE: BRIEF THOUGHTS

Leading from these thoughts about Eichberg's connection with a recast sports geography, and more specifically from the suggestions about body culture and configurations of bodies and landscapes, it is possible to suggest further parallels between Eichberg's work and studies tackling the intersections of body culture with the axes of space and (individual and collective) identity. There are various currents of inquiry in history, anthropology and sociology now talking about the insertion of the human body into social life, recognizing the complex ways in which what this body should be, do and look like are 'constructed' by diverse discourses and practices (Frank, 1989; Turner, 1984). In the literature of academic geography, there are also signs of such an interest as bound into a consideration of how spatial relations – the spaces in and through which bodies move, display themselves and are disciplined – enter into the articulation of bodily presences with the operations of wider socio-cultural formations. There are several different routes by which geographers have arrived at a sensitivity to 'bodily' or 'embodied geographies', but perhaps the most important is the growing literature on the gendering and the sexing of the body as itself a space colliding with spaces beyond the surfaces of skin and clothing. An early statement in this respect was Louise Johnson's 1989 call for geographers to take seriously 'the sexed body in space', in which she developed an example showing how the restructuring of social and authority relations in an Australian textile mill was bound up with 'the mobilisation and redefinition of women's places and bodies' (Johnson, 1989, p. 136). Furthermore, Gillian Rose (e.g. 1991, 1993) has offered similar observations on how 'masculinist' discourses construct women's bodies as certain kinds of entities with certain properties, spatial capabilities and 'proper places' (ones marked by emotions, passions and intimacies), and thereby trap these bodies in discursive fields – drawing out 'bodies as maps of power and identity' (Haraway, 1990, p. 222) – that serve to perpetuate, perhaps most obviously, the public–private divisioning of men's from women's space. What Rose also does is to consider the more experiential dimensions of women's as opposed to men's encounter with space, and in so doing she makes the following telling remarks:

> I'm not quite sure how to specify this difference – only to say that the spaces I feel are women's are very different from the notion of space which time-geography and structuration work with. . . . [A]nd I want to suggest that feminist geographers' accounts of mothers and their time–space zoning, with their stories of childbirth and love, offer a challenge to the strange absence of the body in time–geography. What time–geography traces are paths – bodies become their paths.
>
> (Rose, 1991, p. 160)

In this passage Rose is signalling a larger argument about the reduction of bodies and spaces to the unbending geometries of straight lines and bounded prisms that conceptually (so she argues) underpins much 'masculinist' geographical thinking, in the academy and beyond, and there is here a noteworthy parallel with some of Eichberg's thinking (notably as written through in Chapters 7 and 9 below).

A related trajectory bringing geographers to an awareness of the body lies in the excitement of studies exploring 'the reclamation of other possibilities for a sexualised coporeality through bodily modifications such as piercing, tattooing, scarring' (Pile and Thrift, 1995: see, for instance, Bell *et al.*, 1994; Bell and Valentine, 1995a, 1995b; Cream, 1995), and these are ones that focus on the 'performativity' of the sexed body – of the body expressing its occupant's sexual orientation or deliberate subversions of such orientations – as it moves through the public spaces of streets, clubs and bars (or as it shuns or seeks 'to pass' as 'normal' in spaces where heterosexual codings become overriding: see also Bell, 1995; Valentine, 1993). Then there are a handful of alternative geographies being written about AIDS, ones that challenge the spatial–scientific stress on the mappable geographies (or geometries) of AIDS and insist instead on 'getting closer' to the bodies (to the everyday lives, communities, politics and places) of persons with HIV/ AIDS, their friends and lovers (Brown, 1994, 1995, 1996; Kearns, 1996; Wilton, 1996): not holding them at a distance as mere 'vectors' of transmission, but embracing them as part of a more humane project not afraid of pain, concern and reaching out. And tied in with this claim, it might be noted that at the close of *Geographical Imaginations*, Derek Gregory avows that a task of a 'critical human geography' is one that 'reaches out, *from one body to another*, not in a mood of arrogance, aggression and conquest but in a spirit of humility, understanding and care' (Gregory, 1994, p. 416: emphasis in original).

It has been noted that 'it would be unfortunate if the study of bodily experiences were reduced to sexual politics alone' (Driver, 1996, p. 107), and a related route allowing geographers to discover the body is the recent debate about the character of medical geography. Here Michael Dorn and Glenda Laws (1994) have provided a compelling commentary on the need to widen the optic of the sub-discipline to include attention to the 'medicalising' of certain types of bodies in certain types of places, and in so doing to raise new possibilities for a 'body politics' that launches from the politicisation of embodied and emplaced resistances to the tyranny of controlling social norms in the field of health. More specifically, and obviously informing Dorn and Laws, is an emerging concern for 'disability and space' that examines how the spaces of the body (both as lived and as socially imagined) link up to the spaces of wider environments (notably ones designed by the 'non-disabled' for the 'disabled') (e.g. Hahn, 1986, 1989; and see the recent debate revolving around Golledge, 1993).[2] And yet another instance where geographers are taking into account the confusions of the human body is in relation to methodology, since how the body of the researcher is 'presented' in the scenes of everyday life (Goffman, 1959) can greatly influence the sorts of responses elicited from research subjects, the sorts of information gleaned and the sorts of sites and situations successfully accessed (e.g. Parr, 1995).[3]

This diverse assemblage of works dealing with 'bodily' and 'embodied geographies' proceeds from an equally diverse set of theoretical co-ordinates, although perhaps the most often mentioned are feminism, psychoanalysis, Foucauldianism and phenomenology. Indeed, in her early statement, Johnson put things like this:

> Geography, like all of the social sciences, has been built upon a particu-
> lar conception of the mind and body which sees them as separate,
> apart and acting on each other. An alternate view of the mindbody as a
> unity, socially and historically inscribed, opens the way for a different
> (feminist) geography. Building on psychoanalysis, the historical geog-
> raphy of Foucault and phenomenology can offer a more elaborate
> framework for investigating the sexed body in space which challenges
> existing conceptions of space and time as well as offering a new
> approach to geography.
>
> (Johnson, 1989, p. 137)

There is much in a passage such as this that, we suspect, Eichberg would agree with in principle. Even the gestures to a feminist perspective are ones that we think he would recognise, particularly given clear indications in several of his papers that he supposes a 'sportised', geometric, enclosed sense of space to be associated with a distinctively 'male' version of rationality (see especially Chapter 7), as well as being bound up with 'male' discourses about childbirth, child-rearing and the 'proper' arrangements of the domestic sphere relative to those of the public world (see especially the closing pages of Chapter 3). This being said, he would probably be wary of the post-structuralist turn, which risks draining all substance and vitality out of bodies, couching their importance solely in the realm of representation, in the paper and electronic landscapes of discourse rather than in the immediate concrete and earth landscapes where they breathe, move, laugh and cry.

There is a definite strain of phenomenology running throughout Eichberg's *oeuvre*, then, which means that he is always alert to how real bodies – running, jumping, stumbling, crawling, gazing heavenwards, eating berries from the forest – are in and of themselves bearers of 'knowledges', and are therefore far more profoundly implicated in the making of their worlds than is ever acknowledged by the 'discourse analysts'. In this regard, his views dovetail with those of phenomenological geographers who, particularly during the late 1970s, discussed the 'body-place ballets' or time-space routinised 'habitual body behaviours' through which bodies themselves (largely independent of people's conscious apprehensions) can acquire and act upon their own embodied 'senses of places'. The best-known scholar here was David Seamon (e.g. 1978, 1979, 1980), but leading from a similar founding in the 'body–subject' phenomenology of Maurice Merleau-Ponty, Miriam Helen Hill found that the 'body–world communion' of visually impaired people hints at a 'holistic environmental knowing' in which touching, smelling and hearing all allow the body to 'read' the geography of its immediate surroundings (Hill, 1989; see also Cook, 1991; Rodaway, 1994). Seamon's work has arguably not received the attention that it warrants, and it became all too readily dismissed in the early 1980s backlash against 'humanistic geography' which is anatomised by Steve Pile (1993), but it might be noted that Johnson (1989, esp. pp. 135–6) explicitly retrieves Seamon's efforts as a valuable precursor for a feminist geography of embodiment.

One objection to Seamon's approach is that it universalises the human body, supposing it to possess such a 'deep' phenomenology – such an elemental, almost biological, wiring of its geographical knowing in relation to primitives of space, scale, distance and direction – that it stands 'outside' society, history and (as well) geography: it is untouched by what Eichberg might refer to as the specific socio-cultural–political configurations in which people (and their minds and bodies) operate from day to day. Putting aside whether or not such a criticism of Seamon is totally justified, it is intriguing to trace (as already mentioned) the influence on Eichberg of the Danish sociology of sport, complete with its 'paradigm' of body culture studies in which the starting point for analysis is the axis between body and culture (or, better, between bodies and cultures). As explained in Chapter 7, this paradigm concentrates on the body in conjunction with historical change and cultural variability, seeking to understand the threads that run in every direction possible between situated bodies (the bodies of given peoples in given times and places going about their business, travels, dances, games and sports) and the broader formations (social systems, cultural practices, political organisations) that encompass what people do, think and even feel in these historically and geographically specific situations. And what this paradigm also produces is a style of research that refuses to focus myopically on sport *per se*, but is always striving to see sport in context and as itself an impelling force within the wider social world, in which case the specific body cultures written into specific constellations of sports are viewed as integral to the overarching processes and transformations of a given period and region: German village games casting light on the fragility of medieval Europe; Olympic Games on the institutionalisation of Western modernity; Libyan bedouin games on resistance to the take-up of Western mores, designs and power relations in post-colonial Africa.

Eichberg's stress on body culture does not slip into being a comfortable cultural relativism, though, in which there is no attempt to look beyond particularities to questions, and indeed judgements, about the bigger picture (what one geographer known to us calls 'big picture historical geographies'). For all of the details about oscillations between indoors and outdoors traditions in European sports since medieval times appearing in Chapter 3, for instance, the outlines can still be detected of a powerful narrative about the increasing tendency for sports to be subjected to a geometric-enclosing impulse, one removing bodily activities such as kicking pigs' bladders from the open wastes between villages to the disciplined environment of the football stadium (see also Bale, 1994). In Chapter 9 Eichberg extends this theme, elaborating a remarkable thesis about the 'sportisation' of older games that has increasingly subjected them to the fierce temporal–spatial disciplines of measurement and record-breaking, in the course of which time–space becomes minutely calibrated by the technologies of 'stopwatch and horizontal bar' (see also Eichberg, 1982) and the peculiarities of the natural environment become smoothed out in the uniform geometries of the gymnasium, running track, sports hall and stadium. Furthermore, among several other related claims, Eichberg adds that in the process of the shift from the 'labyrinth' (as an older, curly and often quite irregular site of running, dancing and celebrations: see also Eichberg, 1989, 1990b) to the stadium and its straightened, right-angled, sealed-up and segmented counterparts, so the very understandings of time and space are transformed. He therefore offers the strong claim that modernity's prevailing apprehensions of time–space have themselves been influenced, possibly far more than anyone else has ever acknowledged, by the sportisation impulse that has in effect transformed the games of early Europe – and is arguably now endeavouring to transform the games of places beyond the West – into the temporally and spatially regimented competitive sports that he supposes to be as typical of modernity as are Le Corbusier's slab tower blocks. Yet he does also anticipate new versions of body culture that are perhaps now appearing, maybe drawing inspiration from both premodern European forms and non-Western 'indigenous' ones, and so the outlines of his 'big picture historical geography' are here being muddied by what he sees as a 'postmodern' and perhaps 'hybrid' re-jumbling up of bodies, cultures, times and spaces (see Chapters 6, 8 and 9).

Nonetheless, the bold contours of what he in effect claims about the 'geometricisation' of the body – the subjecting of the body to rigid temporal and spatial disciplines designed to oust ambiguity, play, wilfulness, humour from the sporting body culture of modernity – should strike a chord with a number of geographers who have envisaged a similar historical triumph of an 'iron cage' time–space order over messy disorder (see, in particular, the claims about the 'purification' of space present in Sibley, 1988, 1995). There are undoubtedly flickers of such a vision in Rose's critique of 'masculinist' senses of space built into the geometries of spatial science, time–geography and structuration theory, as inherited from the geometers of the Enlightenment, but so too can such a vision be found in Gregory's attempt 'to connect the history of the body with the history of space' (Gregory, 1994, p. 416; see also Gregory, 1995). Gregory's project draws in particular from Henri Lefebvre's history of urbanism (Lefebvre, 1991), and at one point he speaks of 'the violence of abstraction and the decorporealisation of space' (Gregory, 1994, p. 382), while at another he writes as follows:

> Lefebvre argues that space, which was originally known, marked and produced through all the senses – taste, smell, touch, sound and sight – and which was, in all these ways, in intimate conjunction with the 'intelligence of the body', comes to be constituted as a purely visual field. ... This collective – and historical – passage marks the transformation from absolute into abstract space: 'By the time this process is complete, space has no social existence independently of an intense, aggressive and repressive visualisation. It is thus – not symbolically but in fact – a purely visual space. This rise of the visual realm entails a series of substitutions and displacements by means of which it overwhelms the whole body and usurps its role.'
> (Gregory, 1995, pp. 33–4; quoting Lefebvre, 1991, p. 286)

Gregory is here underlining the importance of an 'occularcentricism' that comes to capture the world in tightly constrained spatial grids of recognition and display, and he is also alluding to a parallel claim about how the 'geometricisation' of knowledge bound up with the Renaissance's (re)discovery of the detached 'perspectival' gaze nullified the importance of the human body in conceptualisations of the world (see Gregory, 1994, esp. p. 389, drawing upon Cosgrove, 1985). These are of course complex materials, and are set within a challenging three-way reading of Lefebvre, Lacan and Harvey, but enough should have still have been said to indicate that Eichberg's narrative of older, sensual body-cultures (full of fun and games, romping about in dirty places) being 'geometricised' by the sanitised body

13

culture of modernity (ruled by the abstractions of quantified time and space) might valuably be considered as occupying the same terrain of concern.

There are certainly other sets of connections that might be drawn out between Eichberg's *oeuvre* and the geographical literature, and most obvious perhaps would be to discuss parallels between what he says about the moulding of bodies through the manipulation of time–space and the borrowings that geographers have made from Foucault on 'panopticism' and the micro-manipulations of time and space in the 'disciplining' of human bodies (Foucault, 1976: see Crush, 1994; Driver, 1985, 1993, 1994; Ogborn, 1991; Robinson, 1990, 1996). This parallel is actually not difficult to observe in Eichberg's writings, not least because he quite often draws explicitly on Foucault's work, and it has already been briefly explored with reference to sport and bodily discipline in the historical geography of 'madness' and asylums (Philo, 1994, esp. pp. 11–14). A more obscure connection maybe arises with a paper by Nigel Thrift (1994), in which he examines the amazing realignments of people, nature and technology occurring in today's 'networks of actants' (thus combining the insights of Haraway and Latour), and where he also charts the shifting historical geography of human–machinic spaces (as transformed in the realms of light, speed and power) that has led to the late twentieth-century fixation on 'mobility' both in and beyond the academy. Pivotal to this account is the human body, since Thrift reasons that the rapidly changing senses of time and space made possible by differing engagements with technology feed into people building up very different 'structures of feeling' and self-identities in response, so that (for instance) the mental constructs associated with viewing a cluttered landscape from the slowly moving 'platform' of a steam train bear scant resemblance to those associated with viewing a map-like landscape from the window of a jet airliner. The differing bodily engagements with the world enabled by such differing forms of movement around and across that world, let alone as now being opened up through the (wholly disembodied?) engagement allowed by electronic media and the access to virtual spaces through the computer screen, are here regarded as foundational of what people conceive of as time, space, community, society, nation, state, past, present, future.[4] There is surely great potential in this respect for Eichberg's notion of body culture to illuminate aspects of these changes and differences, and we can already see stimulating links between Thrift's provocative paper and what Eichberg argues in Chapter 9 about the bedazzling nexus of time, space, identity, modernity and postmodernity.

OVERVIEW OF CHAPTERS

In the chapters that follow we hope to alert readers in fields across the social sciences and humanities to the work of a scholar who provides fascinating insights and applies provocative ideas to body culture and society, space and place. Being conscious of our own 'particular' reading of Eichberg's work, we felt it useful to include the more general essay situating the efforts of Eichberg the 'man' and the 'scholar' more widely; hence the contribution from Susan Brownell in the second chapter. This provides a background to Eichberg's work by focusing on his influential German-language studies, *Der Weg des Sports in die industrielle Zivilisation* (1973) and *Leistung, Spannung, Geschwindigkeit* (1978). Brownell identifies the key contributions and influences of these and other studies. She also alludes to Eichberg's background and the nature of the academic controversies surrounding him in the nation of his birth, Germany. We feel that it is important to recognise the controversial aspects of Eichberg's life and work, and Brownell incorporates an overview of his own feelings about those who have attacked him on political grounds.

We have then arranged the seven papers making up our own selection of Eichberg's writings into three broad groups. Thus Chapters 3 and 4 are concerned with **the body in space**. These two papers have been chosen to illustrate aspects of what Eichberg claims about the 'territorialisation' of the human body. The first (Chapter 3) is an extensive overview that shows how the spatial confinement of the body in 'sports space' can be related to a number of other tendencies that are claimed to have formed a 'great confinement' in eighteenth- and nineteenth-century European society (Foucault, 1967). Here Eichberg provides many sporting examples to illustrate the ideas of a number of scholars who have addressed the disciplining of the body in space (Foucault, 1976; Elias, 1982; Sack, 1986). Sport is a world of disciplined bodies, and scholars dealing with sports at all spatial scales clearly have much to savour from Eichberg's eclectic exemplification of such disciplining in action. At the same time, he also notes a number of 'green waves' in the relationship between the body and the physical environment. This is typical of Eichberg's thinking, since he repeatedly effects a rejection of simple unilinear ways of thinking about history, evolution and 'progress', and it also reflects his 'green' interests that would appear to place himself near the 'deep ecology' end of the environmentalism spectrum. He is careful to note other contemporaneous tendencies in society that are reflected in, or are themselves reflecting, the world of sports. Moving from the 'confining tendencies' identified in Chapter 3, Chapter 4 recognises 'new configurations' that Eichberg sees appearing in the environment of sports. Eichberg observes a 'softening' of the hard configurations of modern sports space, and here he points to ecological and feminist architecture, a return to the open air and a growth in the mixed use of sport space. Although most of the examples found in this essay are taken from Denmark, his adopted home, the general theme of variety and change in the 'landscape' of sports will resonate with readers in most other Western (and increasingly non-Western) countries in the world.

The next two essays, Chapters 5 and 6, tackle **bodies, cultures and identities,** and thereby relate body cultural practices to national or cultural

15

identities. In other words, they reflect Eichberg's interest in the potential for place-to-place differences in body cultural practice to survive in an increasingly homogenised (Olympian) world. In Chapter 5 Eichberg first addresses the adoption of Western sport in Libya, in which he highlights many of the contradictions of a modern Islamic state. Modern sports are found to coexist with traditional Libyan forms of body culture and bedouin games. This contradiction is reflected in Libyan sports policy, which juxtaposes 'Western' commercialism and giant concrete stadiums with 'public' welfare sport, justified as a 'pyramid' after the style of the former German Democratic Republic. At the same time, and again implying 'trialectic' reasoning, there are the traditional bedouin games. The inapplicability of simple dualisms in sports are then related to the problems of an 'either–or' approach to Libyan body culture and society *per se*. Chapter 6 is a short but stimulating view of Olympism and its alternatives. Although this paper now shows some signs of datedness, the often taken-for-granted characteristics of Olympic sports are exposed by Eichberg's incisive critique of their analogies in the excesses of Western capitalism, and at the same time he stresses that 'Olympian' forms of body culture are not seen as 'natural' or necessarily appropriate for all peoples and places. If new anti-colonial movements emerge in the name of 'cultural identity', what does this mean for sport? How can countries of the so-called 'Third World' meaningfully compete with countries of the 'West' in sports like yachting, for example? Eichberg argues that in four areas of body culture alternatives are developing: these are national cultural games, the open-air movement, expressive activities, and meditative exercises. These alternatives pervade the final chapters of the book.

The book concludes with three papers that we identify as pointing **towards a new paradigm** in (at least) studies of the sporting human body and, by extension, of the sporting cultural landscape too. Some of the ideas introduced in earlier essays are here developed and refined. In his paper on 'Body culture as paradigm', Chapter 7, Eichberg makes the notion of the 'trialectic' explicit in order to demonstrate that the prevailing notion of sport is only one way in which the moving, physical body can be configured in modernity. Taking his evidence from Danish research, he again recognises tendencies other than those related to the Olympic ideal of *citius, altius, fortius*. These are non-sportised forms of body discipline arising through physical education and 'sport for all', on the one hand, and the less constrained and freer bodily configurations achieved through more experiential forms of body culture such as fun running, on the other. The implications of such body cultural change has obvious implications for architecture and landscape, as outlined in Eichberg's earlier contribution (Chapter 4) on 'alternative planning' for sport. Chapter 8, drawing on many rich veins of historical detail, identifies a wide range of emerging activities that flesh out Eichberg's earlier assertion about new forms of body culture. As well as describing the ways in which games became modernised, Eichberg again stresses that achievement sport is only one form of that modernisaton, and that other body cultural forms such as folkloric or 'museumised' sports and welfare sports remain central to the workings of the modern world. And then there is also the huge number of heterogeneous forms of movement culture contributing to the cultural identities of minority groups within nation-states. The final chapter, Chapter 9, is a wide-ranging study of time and space in a sports context. Time, argues Eichberg, becomes an 'arrow' of measurement that makes human achievement objective, and space becomes the framework within which the functional needs of achievement-oriented production are fitted. It is easy, given this background, to see sport as a metaphor for the rationality of modern life. Or is it? Eichberg proceeds to argue that current innovations in sports are questioning this image.

Before leaving the stage for first Susan Brownell and then Henning Eichberg himself, a final note should be appended here about the editing of Eichberg's chapters in this collection. Although the original papers by Eichberg were written in English (either directly by Eichberg or in translation from German), we have felt it appropriate to edit them quite heavily so as to make them read more smoothly, and hopefully to allow the sophistication of their arguments to be expressed more clearly than was arguably the case prior to such editing. In every case, we have given Eichberg the opportunity to look over the text and to suggest further revisions, which have now been incorporated into the chapters. There was considerable variation between the original papers in terms of their formatting and styles of referencing, and we have sought to rework them here to give the chapters that follow a standard format and style of referencing (and in the latter case this has meant converting several papers from an endnote or footnote style to the Harvard style). Unevenness in the level of detail included in references has created difficulties, and in the chapters we have had little choice but to go for minimal detail, thus missing out the information that was included in some of the original papers. This has involved the omission of the names of publishers, since these were excluded from some of the original publications. We realise that on occasion this may prove irritating to readers, given that a few references are now probably too abbreviated to be easily followed up. We hope that the resulting product will nonetheless be acceptable, and that it will inspire greater interest in the contribution that Eichberg can make to Anglo-American research on questions of space, identity, body cultures, games and sports.

NOTES

1 In this context it is worth noting the paucity of geographical analyses based on Eliasian ideas (although see Ogborn, 1991).
2 It is intriguing in this connection to think too of what is involved in the equation of 'ability and space', the assumptions about what individual people (and their bodies) are *ideally* supposed to be able to cope with in the external world, and to note – with Hahn (1986) – that so many built environments appear to be produced not even for

the ordinarily 'abled' but for the 'super-abled'. Many city environments are ones that can only be easily negotiated by the fittest, healthiest, strongest and most athletic of people, for instance, and so are hostile not only to those with obvious physical impairments but to very many other people as well. Consideration might be given here to how the figure of the elite sports person feeds into the idealised, even unrealistic, image of bodily capabilities underlying much environmental design and planning, and hence into the complex intersections of ability, disability and spatial forms 'on the ground'.

3 Human bodies are physically marked as different from one another in many ways, of course, and nowhere more obviously than in the case of 'colour' or skin pigmentation. Much geographical research has considered the relations between 'race and space', and by implication has often examined spatial divisions that appear between peoples of different colour as a result of responses prompted by these different colourings of the body (i.e. by racial assumptions and prejudices built around this so-visible bodily mark of difference). Such issues have perhaps been most starkly exposed in the work of geographers researching the Apartheid era and its antecedents in South Africa (e.g. Robinson, 1996; Western, 1981).

4 We are here reminded of the comments made by the geographer Nigel Thrift at a conference in Glasgow on 'Geographies of Power and Resistance' in 1996 at which he alluded to the significance of 'dance' – or, at least, certain kinds (Eichbergian 'configurations') of dancing. Our feelings are that he may have been alluding to the way in which the self becomes transformed through the unusual bodily movements in time–space demanded by dance, a kind of 'losing' of the normal self and a 'gaining' of a new, if fleeting, self totally immersed in the dance, in the moment, in the here-and-now, in the immediacy, in the unmediated encounter between person and world. Such a self, we feel Thrift was arguing, can briefly step outside of conventional power relations, the usual relations of domination and resistance, in which it is embroiled – although an alternative construction would be to see this dancing as itself resistant, as a space of resistance to the routine inscriptions of power.

BIBLIOGRAPHY

Baade, R. 'Stadiums, professional sports, and city economies. An analysis of the United States experience', in Bale, J. and Moen, O. (eds) *The Stadium and the City* (Keele, 1995), pp. 277–94

Bale, J. *Sport, Space and the City* (London, 1993)

Bale, J. 'Cartographic fetishism to geographical humanism. Some general features of a geography of sport', *Innovation in Social Sciences Research*, 1992, pp. 71–88

Bale, J. *Landscapes of Modern Sport* (Leicester, 1994)

Bale, J. 'Rhetorical modes, imaginative geographies and body culture in early twentieth-century Rwanda', *Area*, 1996, pp. 289–97

Bale, J. and Sang, J. *Kenyan Running: Movement Culture, Geography and Global Change* (London, 1996)

Bell, D. 'Pleasure and danger. The paradoxical spaces of sexual citizenship', *Political Geography*, 1995, pp. 139–53

Bell, D. and Valentine, G. 'The sexed self', in Pile, S. and Thrift, N. (eds) *Mapping the Subject. Geographies of Cultural Transformation* (London, 1995a), pp. 143–57

Bell, D. and Valentine, G. (eds) *Mapping Desire. Geographies of Sexualities* (London, 1995b)

Bell, D. *et al.* 'All hyped up and no place to go', *Gender, Place and Culture*, 1994, pp. 31–48

Brown, M. 'The work of city politics. Citizenship through employment in the local response to AIDS', *Environment and Planning A*, 1994, pp. 873–94

Brown, M. 'Ironies of distance. An ongoing critique of the geographies of AIDS', *Environment and Planning D: Society and Space*, 1995, pp. 159–84

Brown, M. 'Time–space and recent historical geographies of AIDS', submitted to *Annals of the Association of American Geographers*, 1996 (date of MS)

Cook, I. *Drowning in See-World? Critical Ethnographies of Blindness*, unpublished MA thesis (Kentucky, 1991)

Cosgrove, D. 'Prospect, perspective and the evolution of the landscape idea', *Transactions of the Institute of British Geographers*, 1985, pp. 45–62

Cosgrove, D. and Rogers, A. 'Territory, locality and place', in Philo, C. (comp.) *New Word, New Worlds. Reconceptualising Social and Cultural Geography* (Lampeter, 1991), pp. 36–8

Cream, J. 'Women on trial', in Pile, S. and Thrift, N. (eds) *Mapping the Subject. Geographies of Cultural Transformation* (London, 1995), pp. 158–69

Crush, J. 'Scripting the compound: power and space in the South African mining industry', *Environment and Planning D: Society and Space*, 1994, pp. 301–24

Digel, H. 'Review of H. Eichberg, *Leistungsräume: Sport als Umweltproblem*', in *International Review for the Sociology of Sport*, 1991, pp. 69–81

Dorn, M. and Laws, G. 'Social theory, body politics and medical geography', *Professional Geographer*, 1994, pp. 106–10

Driver, F. 'Power, space and the body. A critical assessment of Foucault's *Discipline and Punish*', *Environment and Planning D: Society and Space*, 1985, pp. 425–46

Driver, F. *Power and Pauperism. The Workhouse System, 1834–84* (Cambridge, 1993)

Driver, F. 'Bodies in space. Foucault's account of disciplinary power', in Jones, C. and Porter, R. (eds) *Reassessing Foucault. Power, Medicine and the Body* (London, 1994), pp.113–31

Driver, F. 'Histories of the present: the history and philosophy of geography, part III', *Progress in Human Geography*, 1996, pp. 100–9

Dunning, E. and Rojek, C. (eds) *Sport and Leisure in the Civilizing Process. Critique and Counter-Critique* (Toronto, 1992)

Dunning, E., Maguire, J. and Pearton, R. (eds) *The Sports Process. A Comparative and Developmental Approach* (Champaign, Ill., 1993)

Ehn, B. 'National feeling in sport. The case of Sweden', *Ethnologia Europea*, 1989, pp. 57–66

Eichberg, H. *Der Weg des Sports in die industrielle Zivilisation* (Baden-Baden,1973)

Eichberg, H. *Militär und Technik: Schwedenfestungen des 17. Jahrhunderts in den Herzogtümern Bremen und Verden* (Düsseldorf, 1976)

Eichberg, H. *Leistung, Spannung, Geschwindigkeit* (Stuttgart, 1978)

Eichberg, H. 'Stopwatch, horizontal bar, gymnasium. The technologizing of sports in the eighteenth and early-nineteenth centuries', *Journal of the Philosophy of Sport*, 1982, pp. 43–59

Eichberg, H. *Leistungsräume. Sport als Umweltproblem* (Münster, 1988)

Eichberg, H. 'The labyrinth', *Scandinvian Journal of Sports Science*, 1989, pp. 43–57

Eichberg, H. 'Forward race and the laughter of Pygmies. On Olympic sport', in Teich, M. and Porter, R. (eds) *Fin de Siècle and its Legacy* (Cambridge, 1990a), pp. 115–31

Eichberg, H. 'Race-track and labyrinth. The space of physical culture in Berlin', *Journal of Sport History*, 1990b, pp. 245–60

Eichberg, H. (ed.) *Schools for Life* (Copenhagen, 1992)

Eichberg, H. and Jespersen, E. *De grønne bolger* (Slagelse, 1985)

Eichberg, H., Dultz, M., Gadberry, G. and Rühle, G. *Massenspiele. NS-Thingspiel, Arbeiterweihespiel und Olympisches Zeremoniell* (Stuttgart, 1977)

Elias, N. *The Civilizing Process* (Oxford, 1982)

Foucault, M. *Madness and Civilization* (London, 1967)

Foucault, M. *Discipline and Punish* (Harmondsworth, 1976)

Frank, A. W. 'Bringing bodies back in. A decade review', *Theory, Culture and Society*, 1989, pp. 131–62

Galtung, J. 'Sport and international understanding. Sport as a carrier of deep culture and structure', in Ilmarinen, I. (ed.) *Sport and International Understanding* (Berlin, 1984)

Goffman, E. *The Presentation of Self in Everyday Life* (New York, 1959)

Golledge, R. 'Geography and the disabled. A survey with special reference to vision-impaired and blind populations', *Transactions of the Institute of British Geographers*, 1993, pp. 63–85

Gregory, D. *Geographical Imaginations* (Oxford, 1994)

Gregory, D. 'Lefebvre, Lacan and the production of space', in Benko, G. B. and Strohmayer, U. (eds) *Geography, History and Social Sciences* (Dordrecht, 1995), pp. 15–44

Gruneau, R. 'The critique of sport in modernity. Theorizing power, culture and the politics of the body', in Dunning, E., Maguire, J. and Pearton, R. (eds) *The Sports Process. A Comparative and Developmental Approach* (Champaign, Ill., 1993), pp. 85–109

Guttmann, A. *From Ritual to Record* (New York, 1977)

Guttmann, A. *A Whole New Ball Game. An Interpretation of American Sports* (Chapel Hill, N.C., 1988)

Hahn, H. 'Disability and the urban environment. A perspective on Los Angeles', *Environment and Planning D: Society and Space*, 1986, pp. 273–88

Hahn, H. 'Disability and the reproduction of bodily images. The dynamics of human appearances', in Wolch, J. and Dear, M. (eds) *The Power of Geography How Territory Shapes Social Life* (London, 1989), pp. 370–88

Haraway, D. 'A manifesto for cyborgs', in Nicholson, L. (ed.) *Feminism/Postmodernism* (London, 1990), pp. 190–233

Harvey, D. *The Condition of Postmodernity* (Oxford, 1989)

Hill, M. 'Bound to the environment. Towards a phenomenology of sightlessness', in Seamon, D. and Mugerauer, R. (eds) *Dwelling, Place and Environment. Towards a Phenomenology of Person and World* (New York, 1989), pp. 99–112

Jarvie, G. and Maguire, J. *Sport and Leisure in Social Thought* (London, 1994)

Johnson, L. 'Embodying geography. Some implications of considering the sexed body in space', in Welch, R. (ed.) *Geography in Action. Proceedings of Fiftieth New Zealand Geography Conference* (Dunedin, 1989), pp.134–8

Kearns, R. 'AIDS and medical geography. Embracing the other', *Progress in Human Geography*, 1996, pp. 123–31

Lefebvre, H. *The Production of Space* (Oxford, 1991)

Mandell, R. *Sport A Cultural History* (New York, 1984)

Ogborn, M. 'Can you figure it out? Norbert Elias's theory of the self', in Philo, C. (comp.) *New Word, New Worlds. Reconceptualising Social and Cultural Geography* (Lampeter, 1991), pp. 78–87

Ogborn, M. 'Discipline, government and law. Separate confinement in the prisons of England and Wales, 1830–77', *Transactions of the Institute of British Geographers*, 1995, pp. 295–311

Parr, H. '"There's method in this madness". Politics and positionality in mental health research' (1995, MS)

Philo, C. '"Enough to drive one mad". The organisation of space in nineteenth-century lunatic asylums', in Wolch, J. and Dear, M. (eds) *The Power of Geography. How Territory Shapes Social Life* (London, 1989), pp. 258–90

Philo, C. 'In the same ballpark? Looking in on the new sports geography', in Bale, J. (ed.) *Community, Landscape and Identity. Horizons in a Geography of Sport. University of Keele, Department of Geography, Occasional Paper No.20* (Keele, 1994), pp. 1–18

Pile, S. 'Human agency and human geography revisited. A critique of "new models" of the self', *Transactions of the Institute of British Geographers*, 1993, pp. 122–39

Pile, S. and Thrift, N. (eds) *Mapping the Subject. Geographies of Cultural Transformation* (London, 1995)

Pred, A. 'Production, family and free time projects. A time–geographic perspective on the individual and societal change in nineteenth-century US cities', *Journal of Historical Geography*, 1981, pp. 3–36

Pred, A. *Recognizing European Modernities* (London, 1995)

Raitz, K. (ed.) *The Theater of Sport* (Baltimore, Md., 1995)

Robinson, J. '"A perfect system of control". State power and "native locations" in South Africa', *Environment and Planning D: Society and Space*, 1990, pp. 135–62

Robinson, J. *The Power of Apartheid: State, Power and Space in South African Cities* (Oxford, 1996)

Rodaway, P. *Sensuous Geographies. Body, Sense and Place* (London, 1994)

Rooney, J. and Pillsbury, R. *Atlas of American Sport* (New York, 1992)

Rose, G. 'On being ambivalent. Women and feminisms in geography', in Philo, C. (comp.) *New Word, New Worlds. Reconceptualising Social and Cultural Geography* (Lampeter, 1991), pp. 156–63

Rose, G. *Feminism and Geography. The Limits of Geographical Knowledge* (Cambridge, 1993)

Sack, R. *Human Territoriality* (Cambridge, 1986)

Seamon, D. 'Goethe's approach to the natural world. Implications for environmental theory and education', in Ley, D. and Samuels, M. (eds) *Humanistic Geography. Prospects and Problems* (London, 1978), pp. 238–50

Seamon, D. *A Geography of the Lifeworld. Movement, Rest and Encounter* (London, 1979)

Seamon, D. 'Body–Subject, time–space routines and place ballets', in Buttimer, A. and Seamon, D. (eds) *The Human Experience of Space and Place* (London, 1980), pp. 148–65

Sibley, D. 'Purification of space', *Environment and Planning D: Society and Space*, 1988, pp. 409–21

Sibley, D. *Geographies of Exclusion: Society and Difference in the West* (London, 1995)

Soja, E. *Postmodern Geographies: The Reassertion of Space in Social Theory* (London, 1989)

Springwood, C. *Cooperstown to Dyersville. A Geography of Baseball Nostalgia* (Boulder, Colo., 1996)

Thrift, N. 'Inhuman geographies. Landscapes of speed, light and power', in Cloke, P. *et al. Writing the Rural: Five Cultural Geographies* (London, 1994), pp. 191–248 (reprinted in Thrift, N. *Spatial Formations* (London, 1996), pp. 257–310)

Turner, B. *The Body and Society: Explorations in Social Theory* (Oxford, 1984)

Valentine, G. '(Hetero)sexing space. Lesbian perceptions and experiences of everyday spaces', *Environment and Planning D: Society and Space*, 1993, pp. 395–413

Western, J. *Outcast Cape Town* (London, 1981)

Wilton, R. 'Diminished worlds? The geography of everyday life with AIDS', *Health and Place*, 1996, pp. 69–83

2

THINKING DANGEROUSLY
The person and his ideas
Susan Brownell

INTRODUCTION

Although he has influenced a number of influential sports historians – in particular, Allen Guttmann and Richard Mandell – Henning Eichberg is little known in the Anglophone world, and he has also been maligned in his former home country, Germany, for his politics. Both situations share a common ground: Eichberg is an eclectic and original thinker who blazes his own trails, ones not easily followed by others. Politically, his shift from the West German right wing to the left wing in the mid-1970s left many people confused. Intellectually, his engagement primarily with history, anthropology, sociology and philosophy, and secondarily with ethology, art history, architectural studies, geography and ecology, has meant that, since he cannot be pigeon-holed into any one discipline, more traditional scholars from those disciplines have not fully appreciated his work. As a person and a scholar, Eichberg is a liminal figure: he operates in the gaps between political, disciplinary and national boundaries. His liminality is mirrored in a recurring theme in his political and theoretical positioning: he is forever seeking to find the 'third position', the 'trialectic' – or, as he puts it, he is always attempting to systematically seat himself between chairs (Eichberg, 1973, p. 172, 1990b, p. 237). One might suspect that in fact his own liminality compels him to seek out interstices instead of structures. Compounding this complexity, his two germinal early books, *Der Weg des Sports in die industrielle Zivilisation* (*Sport on Its Way into Industrial Civilization*, 1973) and *Leistung, Spannung, Geschwindigkeit* (*Achievement, Tension, Speed*, 1978) have yet to be translated into English. His list of publications since 1970 numbers almost 200, but most are in German and Danish, and only around thirty articles have appeared in English.

And yet his work has still managed to reach out and seize a few Anglophone scholars in the most surprising of ways. Allen Guttmann became acquainted with Eichberg's early work while teaching in Tübingen in 1977–8. *Der Weg des Sports* and two essays were a key source of inspiration for *From Ritual to Record. The Nature of Modern Sports* (Guttmann, 1978), but Guttmann had not read *Leistung, Spannung, Geschwindigkeit* at this time because it had not yet appeared in print (Guttmann, personal communication, 1995). Richard Mandell, who did research for *Sport. A Cultural History* (Mandell, 1984) at the Sport Science Institute of the University of Bonn and at the German Institute of Sport Science in Cologne, also acknowledges his intellectual debt to Eichberg. Like Guttmann, Mandell calls Eichberg 'brilliant' and 'original' (Mandell, 1984, pp. 281, 309). Independently of these two, I became acquainted with Eichberg's work at the German Institute of Sport Science in 1988, when Dr Dietrich Quanz kindly gave me copies of several key articles and loaned me *Der Weg des Sports* and *Leistung, Spannung, Geschwindigkeit*. Despite coming from a different disciplinary background – cultural anthropology, not history – I, like Guttmann and Mandell, was immediately struck by the brilliance and originality of Eichberg's work. My book, *Training the Body for China. Sports in the Moral Order of the People's Republic* (1995), also owes an intellectual debt to Eichberg. We three American scholars were all enthralled by Eichberg's early works in sports history, which we had to read in German. John Bale, by contrast, became intrigued by the smattering of available English translations of Eichberg's later works that reflected his turn to spatial and ecological studies (and is reflected in the choices of essays to reproduce in this volume).

The strength of Eichberg's charisma is to some extent reflected in the events that led to my writing of this essay. Allen Guttmann initially wrote a letter to me complimenting the insight that I had shown in a review of *Ritual and Record. Sports Records and Quantification in Pre-Modern Societies* (Carter and Krüger, 1990); this book, which I will discuss in detail below, was John Marshall Carter's and Arnd Krüger's edited collection of responses to Guttmann's earlier *From Ritual to Record*. He noted that I had favourably reviewed the chapter by Eichberg, and commented that Eichberg was a friend of his. I wrote back to thank him and asked for Eichberg's address so that I could personally express my admiration for his work and my interest in attempting a translation of one of his books. Eichberg responded with a friendly letter in which he also gave me John Bale's address as another scholar interested in publishing translations of his work. Guttmann had earlier recommended to Bale that he read Eichberg's work. I contacted Bale and subsequently met him during his trip to the United States in 1992. Later he invited me to the international conference on 'The Stadium and The City' in Gothenburg, Sweden, in 1993, where I finally had the opportunity to meet Eichberg in person.

In 1988, after Eichberg's written work had piqued my interest, I soon discovered that along with the official, published scholarship comes another phenomenon: the rumours. Even among English-speaking scholars who are practically unacquainted with his actual writing, there are those who have heard that: he is a rightist, an extreme nationalist, an anti-Semite, *persona non grata* in Germany, a dangerous character. One had best avoid association with him. When I heard these rumours, I was astounded: how could it be that the person who had written some of the most culturally sensitive accounts of sport that I had read be any of these things? After meeting him in November, 1993 I was still more the non-believer: with his mild manner and his free-wheeling intellect, he struck me as similar to some of the scholars whom I have admired most in my own discipline of cultural anthropology (which, on the whole, is a liberal discipline strongly antithetical to the kind of thinking attributed to Eichberg by the rumours).

The purpose of this chapter is twofold. First, it will summarise Eichberg's own explanation of his political past so that, with this first major introduction of his work in English, vague rumours do not continue to spread without scholars having some information by which to assess their origins. Second, it will turn to what is really the more important task: to discuss Eichberg's two important early books and outline the interdisciplinary influences on his ideas, in order to provide a more complex understanding of his approach to sports history than that gained from reading the brief citations of his work in Guttmann, Mandell and Carter and Krüger. It is hoped that this analysis will clarify some of Eichberg's key concepts, so that other scholars who might wish to use a similar analytical approach will have a better understanding of how to do so. It is my contention that Eichberg provides us with innovative ways of thinking about the movement of bodies in space, and also the position of bodies within human relationships, which can provide conceptual links between body culture and larger social-historical issues. His work deserves to be considered by the growing number of scholars who are writing social histories of sport, human movement and the body.

THINKING DANGEROUSLY: NOTES TOWARD THE BIOGRAPHY OF AN EXILE

In 1989 Frank Teichmann completed a dissertation at the University of Hamburg entitled *Henning Eichberg – National Revolutionary Perspectives in Sports Science. How Political is Sport Science?* (Teichmann, 1991). The dissertation was a vituperative attack on Eichberg for his (imputed) leadership of the New Right-wing movement in Germany. It explained his multiple publications in left-wing journals as an attempt to 'infiltrate' the left and the Greens in a kind of 'conspiracy'. While I have not read the dissertation, K. Sjöblom (no date) and S. Güldenpfennig (1991) have concluded that the dissertation is an unsound piece of scholarship. In a 1990 article in *Stadion*, entitled '"Gefährlich Denken". Über Rationalität und Angst in der Sportwissenschaft' ('"Thinking dangerously". On rationality and fear in sports science'), Eichberg himself responded to Teichmann's accusations and outlined the personal history that had left him open to such attacks. Much of the same history is repeated in 'The enemy inside. Habitus, folk identities – and a controversial political biography' (Eichberg, 1994). Eichberg's accounts of his political past help to explain how he came to be so maligned. The following biography is drawn from these two accounts. Because I am not myself competent to analyse the incidents described – indeed, for an American, the German academic/political culture described here seems almost bizarre – I have merely tried to summarise Eichberg's own view of events.

Henning Eichberg was born in 1942 in Schweidnitz, Silesia, which was occupied a few years later by the Soviets and annexed to Poland. His place of birth is today called Svidnica. His family fled to Saxony, where he lived from 1945 to 1950. There he witnessed the fire-bombing of Dresden, his earliest childhood memory, and experienced the establishment of the German Democratic Republic (GDR). At the age of eight, his family illegally crossed into West Germany and settled in Hamburg, leaving behind relatives in the GDR. 'As a child, therefore, three "homelands" befell me; this would make finding a homeland a long-term problem for me' (Eichberg, 1990b, p. 229). In the late 1950s he began his search for a political home. Attempting to detach himself from his past and become 'Western', he joined the anti-communist right. He belonged to the Christian Democratic Union (CDU) during the period 1965–8 and the New Right from 1968 to 1975. During these years he was the German correspondent for the French right-wing journal *Nouvelle Ecole* and also published in several German journals. Contact with French nationalists and the student revolts of 1968 brought the contradictions between his anti-communism and the call for national reunification into focus: on the one hand, for the anti-communist right, the United States was the main bastion of the free world; on the other hand, American occupation of West Germany had made it the main nuclear base against the East. Eichberg began to see the anti-communist, pro-American position as a hindrance to national reunification when he realised that, in practical terms, it made contact with his communist counterparts impossible. This realisation compelled his move from the Old Right (the CDU) to the New Right.

He began his studies at the University of Hamburg in history, sociology and German language and literature, but followed Albrecht Timm (an 'academic outsider' who was one of the first to study the history of technology within the context of economic and social history) to the Ruhr University of Bochum. There, in 1970, Eichberg completed his doctoral thesis on Swedish fortresses and engineering in North Germany in the seventeenth century. 'Through it, unusual perspectives on the history of everyday life were opened up to me, on the history of science and the social variability of technology. . . . But there was really no place for that in the university within the framework of the established disciplines' (Eichberg, 1990b, p. 229). He then fell in with 'even more of an outsider', August Nitschke, who brought him to the University of Stuttgart in 1971. In 1976 he was awarded the habilitation degree there. At Stuttgart, they developed what they called 'historical behaviour research' (*historische Verhaltensforschung*), which was based on the notion that '[i]t is not only the world of ideas and institutions, or that of systems of production, that varies throughout history, but also and directly the behaviour of humankind itself, its way of seeing and moving, its senses and its bodiliness' (Eichberg, 1990b, p. 229). It was from this research that Eichberg's interest in the history of sport emerged, resulting in *Der Weg des Sports* (1973), *Leistung, Spannung, Geschwindigkeit* (1978), and *Die Veränderung des Sports ist gesellschaftlich* (*The Transformation of Sport is Social*, 1986). Research in West Sumatra among the Minangkabau in 1974–5 and the Mentawai in 1979 also led to his critique of colonialism.

From 1973–9 he participated in a National Revolutionary organisation that tried to establish a 'third position' (note this recurring motif) between right and left, and did so by reconsidering socialist traditions and integrating a critique of capitalism into discussions of the national question. Participants advocated solidarity with ethnic minorities and concerned themselves with the 'threatened peoples' (Eichberg, 1994, p. 102). Eichberg identifies this period as the source of his later political expatriation:

> The dreams of the potential of another society – democratically-based, free from exploitation, with national self-determination and international solidarity – was a generational thing, which overlapped with the political positions. . . . That I actively participated was an important precondition for my scientific creativity. In the field of the history of technology, as in that of sport science, it contributed to the genesis of a new critical school. But everything has its price. Profits without expenses exist only in the fantasy world of growth theorists. For my hard-earned lesson I had to pay with the 'rumours' that meshed with my collaboration in the national revolutionary organisation *Sache des Volkes* [*The People's Cause*] and their journal *Neue Zeit* [*New Time*] in 1973–79, and after that accompanied me on my way. They are the stuff out of which professional exclusion and also Teichmann's dissertation are made.
>
> (Eichberg, 1990b, p. 231)

By 1975, Eichberg had begun his break with the New Right, which he felt was not open to intellectual exploration of the national question or cultural criticism; it had become conservative, statist and anti-socialist. Meanwhile, the New Left had undergone a transformation since 1968; no longer the enemy 'Reds', they had begun to criticise Western consumerist society as well as Soviet state socialism in a way that was not possible in the New Right. Eichberg had engaged in debates with members of the New Left since 1965, although their dogmatic small groups still did not appeal to him. However, by the mid-1970s new forums for discussion had opened up in the New Left, and from 1976 on he wrote for critical left-wing, alternative and Green journals. He became involved in the ecological movement, and in 1979 helped found the Green Party in Baden-Württemberg. This period resulted in publications that revolved around a critique of technology, alternative cultural movements and questions of national identity.

While the association with the New Right was disastrous for his professional reputation, Eichberg acknowledges that his shift from right to left brought personal growth:

> I took out of the failure of the New Right the lesson that the pathos of feasibility itself is problematic. Instead of 'challenging' or 'abolishing' in the self-confidence of one's own *avant-garde* excellence, there is now a concern with thinking things through, with resistance. The subversive discourse of Michel Foucault was an essential inspiration for this. Not to create programs, but to think against the grain – not last of all against oneself. I would call this a new modesty – if it did not itself sound too immodest.
>
> (Eichberg, 1990b, p. 232)

In his own assessment, it was above all his move to the left, his reformulation of leftist discourse and his contribution to the rise of the Greens that made people suspicious. In 1980–1 the West German Ministry of Culture denied Eichberg employment at two universities. In 1982, *Stern* magazine listed him as a 'Red Nazi'. The most extreme paranoid rumours came from leftist writers who made him the spiritual leader of a 'National Revolutionary Conspiracy' that had infiltrated the top of the Social Democratic Party, with Peter Glotz, the former General Secretary of the SPD, being his most prominent agent. Professionally blackballed in Germany, Eichberg accepted the invitation from the Ministry of Culture in Copenhagen to take guest professorships at the Institute for Sport at the University of Odense in 1982, and the Institute for Cultural Sociology of the University of Copenhagen in 1984. In 1982 he was offered a research position at the Gerlev Idrætshøjskole (a people's academy for sports), where he is still employed. In the 1980s he also gave guest lectures in Finland and did research in Libya on body culture. In 1985 he helped form a research network of French, Danish and German scholars who called themselves the *Institut International d'Anthropologie Corporelle*. In Germany, although he was treated as a traitor by the conservative right, reactions to his suspected rightist conspiracy strengthened after the fall of the Berlin Wall in 1989. Lectures at German universities were cancelled, and German colleagues were criticised for associating with him. In 1993, the leaders of the University of Stuttgart refused to award him the title of professor, against the vote of his faculty and the recommendations of the requested reviewers. The press was informed that, because he had distanced himself from his earlier right-wing positions, he was unsuited for a professorship. Summarising his situation as an example of German political culture, Eichberg observes, 'The disappearance of the "red" enemy had in no way removed the need for an inner enemy as an element of German self-understanding' (Eichberg, 1994, p. 105).

In sum, Eichberg's 'Notes toward the biography of an exile' show that his search for the 'third position' amidst the politics of opposition was perceived as subversive and dangerous. At one level, some of the accusations against him seem overblown and paranoid. However, in the grander scheme of things, perhaps this assessment is not so wrong after all: third positions do, after all, have a way of rendering long-cherished dualisms obsolete, and for those who cherish them this can indeed be dangerous.

THE EICHBERG/GUTTMANN/MANDELL HYPOTHESIS

I turn now to the misunderstanding of Eichberg that has characterised previous Anglophone scholarship. His initial widespread introduction to Anglophone scholars was through the works of Guttmann and Mandell, neither one of whom actually wrote much about Eichberg in their early books. Although Mandell writes that Eichberg was a 'major influence' on Guttmann, Guttmann himself hardly acknowledges Eichberg in *From Ritual to Record*. He comments on Hans Lenk's suggestion that achievement sport is closely connected to the scientific-experimental attitudes of the modern West, noting that this idea was further developed by Eichberg in *Der Weg des Sports* (Guttmann, 1978, p. 85). Two pages later, Guttmann notes that he finds Eichberg unpersuasive when he correlates the rise of modern sports with the Romantic revolution of the eighteenth and nineteenth centuries (Guttmann, 1978, p. 87). Those are the only mentions of Eichberg in the text of the book.

Nevertheless, Mandell lumps together Bero Rigauer, Eichberg and Guttmann as critics of sport who argue that sport is a unique adaptation to the modern 'achieving society' (Mandell, 1984, p. 3). Interestingly, Mandell also emphasises the importance of a part of Eichberg's biography that Eichberg's own two accounts entirely omitted. Mandell states that preparations for the 1972 Olympic Games provided an important context for Eichberg's emergence as a pro-ecology, populist critic of elite sports (Mandell, 1984, p. 281). Eichberg begins his foreword to *Der Weg des Sports* (completed in 1972) by mentioning that the budget increase for the Munich Games from DM520 million in 1966 to DM1.972 billion in 1971 called for criticism (Eichberg, 1973, p. 7). Mandell identifies the convergence of the Munich debates and the historical behaviour research at the University of Stuttgart as a moment that inspired Eichberg to write his important 1974 article ' "Auf Zoll und Quintlein". Sport und Quantifizierungsprozeß in der Frühen Neuzeit' ('Towards inches and grams. Sport and the process of quantification in the Early Modern'). This article was later cited by Carter and Krüger as the beginning of the debate over ritual and records in sport, or what they called 'the Eichberg/Guttmann/Mandell hypothesis' (Carter and Krüger, 1990, p. 1). In brief, this hypothesis states that modern sports emerged along with modern industrial society, that both are characterised by an emphasis on 'achievement' as seen in sports records and economic productivity, and that both are qualitatively different from anything that preceded them historically in Western Europe or that was found in non-Western societies before Western contact. All three scholars share the assumption that sports are culturally variable and that they change through time; they also locate the beginning of modern sports in eighteenth-century England, and they identify a high level of rationalisation and quantification as a key defining feature of modern sports (Guttmann, personal communication, 1995). In 1980 the key players in this debate were brought together at a conference at UCLA on 'The Beginning of Modern Sports in the Renaissance', and their articles eventually formed the core of *Ritual and Record*.

In developing his version of the 'origin of records' hypothesis, Mandell used Walter Rostow's theory of economic take-off (Mandell, 1976). Guttmann's version of the hypothesis, which, being book-length, has been the most influential, borrowed heavily on the modernisation theory outlined by Weber and Parsons. In other words, both Mandell and Guttmann fitted sports within the framework of a unilinear concept of the kind of economic modernisation and 'progress' unique to the West. It is, therefore, highly ironic that Eichberg was lumped together with them. All along he has resisted his inclusion in this formula, a fact that has only been briefly dealt with in the sources cited. In his own contribution to *Ritual and Record*, Eichberg forcefully states his view of the entire argument: a history of records is anachronistic because it projects backwards the constructions of record, sport and achievement that are products of industrial culture (Eichberg, 1990a, p. 125). Such a history only makes sense in nineteenth-century England, when movements were used to produce concrete results in the form of numerical figures. At that time, 'legal' conditions for records were defined, space made uniform and controlling organisations formed (Eichberg, 1990a, p. 129). For Eichberg, the quintessential modern sport is track and field, with its emphasis on linear motion, speed and timekeeping; it is no more than 200 years old (Eichberg, 1990a, p. 131). His argument against his inclusion in the formula is hence based on his assertion of the cultural relativity of sports, and on his arguments against de-ritualising and totalising history. Only parts of this argument seem fully justified, as Guttmann and Mandell also allow that sports are culturally relative and historically variable, but they do use 'totalising' varieties of modernisation theory (accounting for the historical changes of the last two centuries) that Eichberg rejects – but it is not clear what Eichberg is offering in the place of modernisation theory. As discussed below, his description of the changes in sports is not conceptually anchored to anything other than general configurational changes taking place in society as a whole, which in his view are following a course of history characterised by discontinuities rather than by cause and effect. Although Eichberg is trying to write an account that is simultaneously a Foucauldian 'anti-history' and a critique of modernity, the differences between his works and those of Guttmann and Mandell are not always crystal clear.

Finally, Eichberg criticises Guttmann's claim that the history of sports reveals a shift from an emphasis on rituals to one on record-keeping. While Guttmann acknowledges that there are modern rituals, he argues that modern sports are secular. Eichberg states that ritual is still alive and well in contemporary sport, and that a history of the 'rationalisation' of records is simply a myth: one that echoes the myth of 'progress' that characterised the modern period (Eichberg, 1990a, p. 132). In fact, Eichberg is fond of calling modern sport 'the ritual of records' and contrasting it with the 'culture of laughter'. This refers to his view that sport changed from a carnivalistic popular cultural form into a record chase regarded with sacred seriousness (Eichberg, 1990a, p. 132), and laughter is a recurring theme in Eichberg's works. It resonates with my own observations in China, where unsophisticated sports fans in the 1980s most appreciated sports performances for their comedic potential: they went to basketball games in order to gawk at unusually tall people; at track meets they congregated by the water jump hoping to watch a steeplechaser take a nose-dive. This contrasts with the sophisticated Western sports fan's concentration on the performance and the results. It is unclear whether the other participants in the ritual and record debates got this point.

Given Eichberg's critical distance from the entire history of the record, why was his work combined with that of Guttmann and Mandell in the first place? The answer to this question is most evident in an analysis of the key book in the debate (and the book that Teichmann's dissertation seized upon as revealing Eichberg's true National Revolutionary leanings), *Der Weg des Sports*. In 1990 Eichberg summarised the intent of this book as follows:

> This book was constructed so that I first sounded out different interpretations of sport to test their strength and durability – sport as game, as biological function, as cult, as profession, as aggression, as pedagogical exercise, as means of national identification, as leisure activity and as work. This 'Indeed' was followed by the 'But': none of these relational systems adequately explained the birth of sport in its specific modern form. For that – so went the main idea – sport must be seen in connection with the process of rationalisation and achievement orientation.
>
> (Eichberg, 1990b, p. 239)

In outlining his approach in the book itself, he stated that:

> Instead of the customary division of sport history into countries, types of sports, or chronology, my focus is on a division according to the

functions of sport in society, or rather according to systems of relationships within which sport can be seen as a social phenomenon
(without implying that this is all that can be said). Among the systems
of relationships observed more or less until now, attention should be
directed toward those whose contribution to the relationship of sport
with the rise of modern industrial society was of special significance.
For this it will be necessary to have at our disposal comparative material
from societies other than the Occident and epochs other than the
modern.

(Eichberg, 1973, p. 10)

In sum, then, Eichberg's main goal in *Der Weg des Sports* was to outline some
of the simultaneous features of the appearance of modern sports and of
industrial society in order to show the uniqueness of modern sports: 'Our
sports are not the games of others' (Eichberg, 1979, p. 174). From this
relativistic point, he could launch critiques of modern 'achievement society',
Olympic universalism, the national question in sport, and the Green or
'ecological question' in sports (Eichberg, 1979, pp.171, 175). Remember that
this was the book written in part as a response to the debates about the
Munich Olympics, but also that it came out just before Eichberg's move to
the left in 1976. As a product of this transitional period in Eichberg's life, the
book contains contradictions that open it up to different interpretations. In
the 1979 postscript to the book, he acknowledged that he had 'followed a still
all-too-naive evolutionism' (Eichberg, 1979, p. 169). By 1990 it seemed that
he wanted to distance himself from the book. He admitted that the book
comes close to recounting a historical transformation of sport from ritual to
record, but that he finds this problematic (Eichberg, 1990a, p. 132). In 'Thinking dangerously' he mentions his 'quarrel with those linear descriptive
models established in sociology, "regularities" and interpretations, lastly
therefore also with my own scientific and political starting points' (Eichberg,
1990b, p. 235):

> Thinking in the categories of (one) rationality, civilisation, achievement-
> orientation, evolution, technical progress, productivity, etc., was shown
> by the history of the body to be the superstructure of a specific mod-
> ern, western–industrial pattern of behaviour. Its claim to universality is
> not upheld by sensible intercultural comparison. Evolutionist and func-
> tionalist categories, the entire construct of sociological universalism,
> prove to be historical, that is, relative.

(Eichberg, 1990b, p. 235)

Furthermore, the linear–progressive theory of modernisation has concrete
implications for non-Western peoples:

> The special dynamic of the Western analytical model is, namely, not to
> think separately of the history of colonial expansion. Where universal

31

regularities are 'recognised,' the *other* of foreign cultures is thought away just as it is levelled away in the colonial reality.

(Eichberg, 1990b, p. 235).

Eichberg repeated a similar point in the foreword to his next book, *Leistung, Spannung, Geschwindigkeit*, stating his desire to keep a sceptical distance between himself and linear–progressive interpretations, 'which see in the past only a step leading to the present, without any potential for the "Other" and the "Foreign" in history (nor in foreign, non-European societies)' (Eichberg, 1978, p. 8). In a sense *Leistung, Spannung, Geschwindigkeit* may have been the book that Eichberg wished he had written the first time around. While it is to some degree a continuation of the first book, in that it examines in more detail the historical source materials for the same time period, it is organised by a more sophisticated conceptual framework that goes further beyond a linear, historical interpretation. Mandell identifies it as the essential work for future scholarship (Mandell, 1990, p. 128). It is also my opinion that this is the key work for understanding Eichberg's thought and subsequent work on ecology. It is useful, therefore, to analyse this work in order to answer the questions: what was Eichberg really trying to say with his sports history, and why would it appear that he was misinterpreted?

'HISTORICAL BEHAVIOUR RESEARCH'

As historians, Guttmann and Mandell were probably unfamiliar with Eichberg's intellectual milieu and the influences that shaped his thought when they first encountered his work. In fact, it was through Eichberg that Guttmann discovered Norbert Elias, who became an important inspiration for Guttmann's later work (Guttmann, personal communication, 1996). I make this statement as an observation, not a criticism, for they should be commended for being two of the few Anglophone scholars to read the rich German scholarship on sports. Further, I do not wish to claim more understanding of Eichberg than they themselves have. What I do wish to do here is to highlight the sociological aspects of Eichberg's work that are glossed over if one reads him purely as a sports historian. Mandell criticises Eichberg for his neglect of English and American sociological literature, while at the same time complaining about his use of sociological jargon borrowed from American sociology (Mandell, 1984, p. 309, 1990, p. 128). Sjöblom, a historian, also complains about Eichberg's sociological jargon (no date, p. 4). My response to these criticisms is threefold. First, I would argue that Eichberg is not strictly a historian: instead, he attempts to combine anthropology and history. Further, he argues that he is writing the kind of 'anti-history' advocated by Foucault. Second, the jargon that historians cannot understand is not borrowed from American sociology: in fact, it is borrowed from the European sociology and philosophy of people like Foucault, Elias and the

Frankfurt school; in addition, there is some influence from American cultural anthropology. If the jargon sounds American, then it is because American sociology has also been influenced by these developments. I also take issue with the way that these two scholars trivialise important insights by calling them 'jargon'. This language is essential to understanding Eichberg's way of thinking and his approach to sports history. Third, as someone who has searched the English and American sociological literature for authors who have had anything important to say about the cultural relativity of sport, I would argue that in the 1970s there was almost no relevant English and American literature for Eichberg to neglect. (Thankfully, this situation is now changing somewhat.)

Before moving on to an analysis of Eichberg's way of thinking, I would like to add one further observation. Previously, I mentioned that, upon our first encounter with Eichberg's work, Guttmann, Mandell and myself were all struck with his originality and brilliance. However, in the seven years since my first encounter, I have become more aware that Eichberg builds on a pool of sophisticated German scholarship on the history of the body and behaviour, almost none of which has been translated into English. While I can in no way claim mastery of this literature, quick glances at the works of people like August Nitschke (1981, 1985, 1987), Rudolf zur Lippe (1988) and Klaus Schreiner and Norbert Schnitzler (1992) have left me tantalised at their (to an English-speaker) novel scholarship. Without taking anything away from Eichberg's own innovative contribution to this literature, I also now see that his originality and brilliance did not emerge in isolation. Adding to Mandell, I would lament that it is not only to the discredit of the sports world that this German movement of the 1970s remains unknown to Anglophones, it is also to the great discredit of the current burgeoning interdisciplinary scholarship on 'the body'.

The resonance between Nitschke's and Eichberg's historical works is, of course, notable. Nitschke cites Eichberg frequently. In his programmatic essay, Nitschke states that the central assumption of 'historical behaviour research' is that there is no single, universal human being who is the same through all historical epochs; thus, the historian should study people relative to the systems of thought and belief that existed in their times.[1] Nitschke calls for an 'ethnological' approach to history, a 'historical anthropology' (Nitschke, 1985, p. 664). In summarising the goal of 'historical behaviour research', Nitschke states:

> Man [sic] always lives in symbiosis with a part of his environment – this is the specialty of anthropology, which is worked out in our Stuttgart Institute for the History of Behaviour. This 'symbiosis' primarily makes effective action possible for him. At the same time it permeates his ideas of time and space.
>
> (Nitschke, 1985, p. 668)

As has been discussed, this central assumption of the relativity of history characterises Eichberg's thinking as well. It is interesting to note, then, that Nitschke criticises *Leistung, Spannung, Geschwindigkeit* and other pre-1980 publications of Eichberg's for not being relative enough. He observes that those sports scientists who have concerned themselves with the achievement principle have been content to assert that in other societies, in 'other' times and places, movement forms are a medium of bodily experience, communication or relations; but they have not then investigated in detail what is meant by this. The footnote to this assertion about unspecified sports scientists cites only Eichberg and one other source, and not Guttmann. Although Nitschke acknowledges that Eichberg's *Sozialverhalten und Regionalentwicklungsplanung (Social Behaviour and Regional Development Planning*, 1981) is the most detailed of these studies, he still criticises Eichberg for giving only general hints, taking European society as his starting point, and failing to do full justice to non-European societies (Nitschke, 1987, pp. 18, 171 n. 9). He criticises Eichberg for using a developmental scheme in which so-called simple, 'natural' human movements develop over time into complex, regulated movements. He questions whether such a 'development' has actually occurred, and complains that it is not at all clear what is meant here by 'natural' (Nitschke, 1987, pp. 20, 172 n. 19). It is clear, then, that even Nitschke did not perceive in Eichberg's work a clear break with unilinear, Eurocentric development schema.

Why was Eichberg's work misinterpreted? Perhaps because it did not go far enough beyond the linear–progressive modernisation theory that it intended to discredit. In 1987 Nitschke could criticise Eichberg for not fully carrying out the relativising task of historical behaviour research, but in the late 1970s this was still a new direction with few models to draw on. The tentative attempt at a new direction and its resulting ambiguity help explain how Eichberg could be lumped together with Mandell and Guttmann in a way that he later tried to reject.

THE CONFIGURATION CONCEPT

In 'Thinking dangerously' Eichberg lists what he sees as the central ideas in his sports analysis: the configuration concept; the experience of space and time; the critique of alienation, identity and Western productivity; and the modernity of sport as a historical leap (Eichberg, 1990b, p. 238). I would argue that the single most important concept to grasp in understanding Eichberg's early sport–historical work – and also its continuity with his later spatial–ecological work – is the concept of configuration. Inspired by the ideas of Ruth Benedict and Elias, Foucault and Nitschke, Eichberg has created a synthesis that goes far beyond its original constituent parts.

In 1934 Benedict translated the German *Gestalt* as 'configuration', and borrowed from Gestalt psychology the notion of starting any analysis from the whole rather than from its parts (Benedict, 1989, p. 51). It is interesting that, although she herself was American, the sources of Benedict's intellectual inspiration were German. In addition to learning from Gestalt psychology, she was a student of the founding father of American anthropology, Franz Boas, a German immigrant. Calling her approach 'configurational anthropology,' she argued that '[a] culture, like an individual, is a more or less consistent pattern of thought and action' (Benedict, 1989, p. 46). Further, she claimed that 'fundamental and distinctive cultural configurations . . . pattern existence and condition the thoughts and emotions of the individuals who participate in those cultures' (Benedict, 1989, p. 55). 'Configurational anthropology' was a way of analysing the intersection of individual psychology and history, both of which were necessary for the interpretation of cultural forms (Benedict, 1989, pp. 232–3). This approach was above all Benedict's response to claims for the biological bases of social phenomena, a claim that she emphatically rejected (Benedict, 1989, pp. 233–4). For Benedict, configurations or 'patterns of culture' should constitute the basic unit of sociological analysis:

> The significant sociological unit, from this point of view, therefore, is not the institution but the cultural configuration. The studies of the family, of primitive economics, or of moral ideas need to be broken up into studies that emphasise the different configurations that in instance after instance have dominated these traits.
>
> (Benedict, 1989, p. 244)

Norbert Elias's concept of 'figuration' covers some of the same ground and, like Benedict's concept of configuration, traces its roots to Germany. As a student in Heidelberg and Frankfurt in the 1920s and 1930s, Elias was motivated to develop a new mode of sociological thinking that could go beyond the entrenched individualism of previous schools:

> What pushed [the early generation of sociologists] towards sociology in many cases was undoubtedly the realisation that in the course of increasing urbanisation and industrialisation on the level of social activity, a profusion of new problems was arising which history, economics and the other social sciences were allowing to lie fallow because they did not fit into their patterns of problem-solving and were not accessible to their traditional methods.
>
> (Elias, 1987, p. 131)

These scholars felt confronted with the 'task of elaborating a comprehensive theory of human society, or, more exactly, a theory of the development of humanity, which could provide an integrating framework of reference for the various specialist social sciences' (Elias, 1987, p. 131). Elias's major two-volume work, *The Civilizing Process* (1978, 1982), was a result of this striving. At once historical and sociological, it attempted to trace the process of state centralisation in sixteenth- to eighteenth-century Europe, and then to sketch its relationship to the transformation of everyday life – particularly manners and etiquette – through the desire of different social classes to represent themselves as more 'civilised' than others. The concept of 'figuration' was coined as a way of placing the 'individual' and 'society' side by side in analysis, with neither getting primacy over the other (Elias, 1987, p. 135). A 'figuration' is 'a web of interdependent people who are bonded to each other on several levels and in diverse ways' (Dunning, 1986, p. 10). 'Figuration' is sometimes interchangeable with words like 'system', 'structure' or 'pattern', but these words had already been used in sociological theory in ways that exemplified the reification of component parts of society, which Elias was trying to counteract with his notion of the whole as a dynamic process that cannot be reduced to the sum of its parts. Figurational analysis is central to any investigation of sports, in part because every sport is an organised group activity that is contested according to known rules that determine the pattern of the contest (Elias, 1986, p. 155). Such figurations are open to historical analysis because they continue to exist even when the individuals forming them have died and been replaced. What remains the same is not the individuals, but the relationships between them; their relationships are finally determined by conditions that are unalterable by the individuals themselves (Elias, 1983, p. 142).

The final source for Eichberg's notion of configuration is the kind of approach that characterised Michel Foucault's 'archaeology' period, especially as found in *The Order of Things* (Foucault, 1970). In this book, Foucault investigates the threshold that separated classical thought from modernity. The object of his analysis is forms of knowledge. In classical thought, he argues, theories of representation, language, the natural orders and wealth and value possessed a coherence among them that he calls a 'configuration'. From the nineteenth century onward, this configuration changed entirely (Foucault, 1970, p. xxiii). Foucault summarised his 'archaeological' approach by stating:

> I am concerned here with observing how a culture experiences the propinquity of things, how it establishes the *tabula* of their relationships and the order by which they must be considered. I am concerned, in short, with a history of resemblance ... the history of the order imposed on things would be the history of the Same – of that which, for a given culture, is both dispersed and related, therefore to be distinguished by kinds and to be collected together into identities.
>
> (Foucault, 1970, p. xxiv)

Foucault's 'archaeology of knowledge' uncovers a 'network of analogies' in the human and natural sciences, an epistemological field, and this network he describes as comprising 'configurations within the space of knowledge that have given rise to the diverse forms of empirical science' (Foucault, 1970,

p. xxii). While he (or at least the English translator) occasionally uses the word 'configuration', the label that he seems to prefer is '*épistème*', although I suspect that 'configuration' and '*épistème*' can be regarded as interchangeable. He defines the latter as the total set of relations that unite discursive practices and give rise to epistemological formations (Foucault, 1972, p. 191). Foucault labels his method 'archaeology', and not 'history', because he is interested in defining systems of simultaneity and revealing discontinuities between them. His focus on discontinuities is precisely the point on which he (and Eichberg) are opposed to the kind of approach taken by Guttmann, Mandell and other participants in the writing of the history of sports records. Unlike traditional ('total') history, which looks for continuities, Foucault seeks for epistemic breaks. He calls for a method that will obtain a plurality of histories, juxtaposed and independent from one another, but characterised by coincidences of dates and analogies of form and meaning. The problem is to determine what form of relation may be legitimately described between these different series (Foucault, 1972, p. 10).

In 'Thinking dangerously' Eichberg notes the importance to him of the insight that 'history makes leaps':

> In the seventies the linearity of the historical process began to dissolve in the analysis of technology as well as the history of the body. In its place another history revealed itself: that of leaps and structural breaks. What had previously been revealed by Foucault as the change in the order of things and knowledge, or by Thomas S. Kuhn as the paradigm shift in the history of science, was suddenly visible at the level of the body and behaviour patterns.
>
> (Eichberg, 1990b, p. 234)

This attention to historical breaks — not a belief in the inevitability of industrial progress — underlies Eichberg's conception of the uniqueness of modern sports:

> Modern sport with its production in c–g–s data (centimetre, gram, second) and its spatial–functional delimitation is not more (and not less) than other historical body cultures: it is an *other*. In its patterns of space, time, energy and bodies — which, following Ruth Benedict, Norbert Elias, Michel Foucault and August Nitschke could be called configurations — the models and myths of industrial society find expression.
>
> (Eichberg, 1990b, p. 234)

Finally, in what I find to be a particularly evocative passage, Eichberg explains how this view of historical processes constitutes the common theme running throughout his interests in the history of technology, the body and sports:

For the history of technology, the new perspective meant that the previous orientation toward innovations, discoveries and popularisation, toward creation and production, became problematic. In technological development not only does the new come into being, but also structures are revolutionised, and with that the old disappears. Thus a history under the aspect of disappearance, of ruins, of (auto)-graveyards is questioned. . . . But nothing disappears completely. Precisely the bodies carry signatures of sunken worlds in themselves and on themselves: tattooing and laughter, body armour and jewellery, illness and rhythms. In the most surprising situations these subterranean bodies manifest themselves, become political and subversive, constitute sub- and counter-cultures. Sport was a product of such historical obstinacy [*Eigen-Sinn*, 'own sense']. The sportive body became, to be sure, productive (data-producing) and streamlined – and indeed not just this. The industrial culture was illustrated here with its achievement, discipline and measurement – and yet, at the same time, came expressions of obscenity, violence and laughter, other worlds of sensuality.

(Eichberg, 1990b, p. 234)

In our discussion in November 1993 Eichberg told me that he considered the configuration concept (*Konfigurationsbegriff*) to be central to his post-Elias, post-Foucault perspective on the body. He suggested that the concept includes five aspects: space, time, interpersonal relationships, objectification and energy. The aspect of time was drawn from Foucault and refers to the ideas outlined above. The aspect of interpersonal relationships was drawn from Elias, and refers to the figurational sociology advocated by him that entails a fight against reification and the 'it' – a point on which Eichberg feels that he is even more radical than Elias. The aspect of objectification is Eichberg's own, and is drawn from his research on quantification and rationalisation. The aspect of energy is a recent addition to the concept, and draws on notions like the German *Kraft* or Chinese *qi* ('strength' or 'energy'); this aspect awaits further development.

While Eichberg said nothing about his concept of space, I think it is worth considering in more detail because I find it to be perhaps his most important contribution. It also forms a pivotal point of continuity for his later ecological works. Eichberg's consideration of space might owe something to Foucault, who showed the disciplinary effects of the arrangement of spaces like prisons, barracks and mental asylums. It might also owe something to Foucault's notion of 'the gaze', as most clearly developed in discussions of the Panopticon, the all-seeing central surveillance point in Bentham's ideal prison. While Elias was quite aware of space (see *The Court Society*, 1983), and was fond of illustrating the figurational concept with spatial diagrams, he did not specifically include space in his conception of figurations. Eichberg goes beyond both of these scholars in that, above all, his work gives the reader a new way to see – to see movement in space, and especially the movement of human bodies in space – and more than that, to be aware of the positioning of 'the gaze' in space relative to all of these.

I find this aspect of his work particularly clear in a brief, seldom-cited article on historical illustrations of sports and dance (Eichberg, 1980). This article summarises the main arguments of *Leistung, Spannung, Geschwindigkeit*, but complements them with illustrations. In following the sequence of his argument, one can see more clearly than elsewhere the influence of Foucault and Elias, and how Eichberg synthesises and ultimately goes beyond both of them:

> From the sixteenth century to the end of the eighteenth, pictures of equitation from the classical era of dressage and the High School show a spatial and position-oriented art of Form. What was taught there in the knightly academies, and by court- and drill-masters, what was described and illustrated in the writings of Grisone (1552), Löhneysen (1609), Pluvinel (1623–5), Newcastle (1658–67) and Guerinere (1773), was an art of placement and circulation in space. The posture of the rider, choreographed hoofbeat patterns, the equine postures of the High School, formed the basic principles. . . . After the mid-eighteenth century there emerged in its place – first in England – representations of stretched-out horses that emphasised the horse 'torn apart' in the horizontal, and thus only the dimension of forward movement. . . . Painting makes the new configuration more than clear: the reduction of the spatial dimension to a line, which is oriented toward a goal, that is, moving along the time axis.
>
> (Eichberg, 1980, pp. 20–1)

In his own innovative way, Eichberg seems to respond to Foucault's call for the historian to search for analogies of form and meaning between disparate configurations when he proceeds to relate these equestrian illustrations to changing conceptions of space and time:

> In the illustration of sport, those patterns also appeared that laid to rest the notion of performance at that time: before 1800, positional movement, geometric regulation of choreography and carriage, a certain tectonics and topography of movement, and rhythmic circulation define the illustration of drills. After 1800, an increase in performance, tension, and thus an emphasis on the dynamics of time, permeated sport. It thus brought along a certain shift from the spatial orientation to a temporal orientation. A finding that refers to an anchoring of changing thought in the elementary presuppositions of behaviour, in social praxis.
>
> (Eichberg, 1980, p. 22)

Next, in a move reminiscent of Elias's analysis of the court society, Eichberg proceeds to link these changing conceptions of time and space to changes in interpersonal relations:

> From the changes developing throughout the medium of illustrations one can gain new insights into the historical conditions of our concept of achievement outside of sport. As the drills before 1800 and sport after 1800 were two configurationally different phenomena, so also were performance (*Leisten*) in the pre-industrial society and the modern concept of achievement (*Leistung*) differentiated; and here also the break lay in the end of the eighteenth and beginning of the nineteenth century.[2] ... The old notion of performance meant: to follow an obligation, to perform or fulfil a debt. Something like, as one still says today, to perform a series of commands, to perform a payment, to perform a service, a loyalty, an obedience, a duty, a help, or an assistance. The performance always received its substantive completion first from an object, from the obligation that it was to fulfil. In the waning eighteenth century, next to this performance another now appears in which first of all the object, the obligation, fades. One achieves 'something', 'much' or 'little'. Then the object could also be completely omitted. One could now, as Goethe differently did, speak of the 'Life, Achievement and Suffering' or the 'Deeds and Achievements' of a person. This new performance possessed not only a new social–historical content, namely the performance and the 'productivity' of the economising citizen and of the 'creative' intellectuals. It possessed also a new structure: instead of the closure of the pre-given and to-be-filled in frame, and a certain measure of choreography, to which the old performance referred, the new performance contained an open temporal horizon, a direction, forward progress, and a climb into infinity.
>
> (Eichberg, 1980, p. 22)

In other words, Eichberg outlines developments in equestrian sports that are part of an epistemic break. This break occurred as one 'network of analogies', or configuration, gave way to another. This break can be traced in changing notions of space, time, interpersonal relations and objectification. Eichberg starts with historical illustrations of dressage and steeplechasing. He then situates the change from positional choreography to a linear race against time within larger changes in the conception of space and time that occurred around 1800. Then he finds an analogy between these changes and the shift from the notion of 'performance' (which characterised interpersonal relationships set in a bounded hierarchy) to 'achievement' (which characterised interpersonal relationships within an open economy oriented towards production). These are all objectified in equestrian illustrations and in the increasing use of the stop-watch to measure achievement after 1730.

40

Voilà – a concise example of how concepts of space and time, interpersonal relationships and objectification constitute a configuration.

The importance of space in the analysis of configurations opens the door for Eichberg's future ecological works, such as *Leistungsräume. Sport als Umweltproblem* (*Achievement Spaces. Sport as an Environmental Problem*, 1988), with its evocative illustrations of parking lots, stadiums, gymnasiums and labyrinths. It has also led to works-in-progress on the nature of 'the gaze' in sports spectatorship: the panoptical view of the stadium, the 'zapping' view of the television set and the 'labyrinthine view' of folk dance and sports.

BODY CULTURE

One final concept that has become associated with Eichberg and other scholars working in Denmark is the concept of 'body culture'. This concept was not prominent in Eichberg's early works, as discussed above, and neither is it prominent in Nitschke's works. It does not seem to become important until the 1980s. Eichberg traces the roots of the phrase back to the German notion of *Körperkultur* (literally, 'body culture'), which first appeared around 1900–5, during the 'free body culture' movement that advocated diet and clothing reform, nudism, sport, gymnastics, folk dance, abstinence from nicotine and alcohol, and so on (Eichberg, 1993, pp. 257–8). These ideas were adopted by the German Socialist Worker's Movement, being anticipated in some respects by Karl Marx, and they eventually crystallised under the label of *Fiskultura* in Russian, which was in turn translated back into English as 'physical culture'. With the Soviet and East German pursuit of gold medals, 'physical culture' became identified with high-level sports – quite the reverse of its original conception in opposition to elitist sports. However, in West Germany and Denmark the corresponding concepts (*Körperkultur*, *kropskultur*) underwent a renaissance after 1968, and have become key words in a new perspective on the body that has been emerging in the last two decades (Eichberg, 1993, p. 258). This perspective looks at the body primarily as cultural; which is to say, as socially constructed and historically variable. The evolution and meanings of this concept are discussed further in 'Body culture as paradigm. The Danish sociology of sport' (Chapter 7). In this article, Eichberg argues that, by means of this concept, Danish sports sociology has distinguished itself from the American–Scandinavian paradigm because it places the body at the centre of the field where it belongs. American–Scandinavian sports sociology, by contrast, is regarded by Eichberg as simply another hyphenated sociology that puts society at the centre of its analysis and relegates the body to the natural sciences.

In my opinion, the concept of body culture is a very useful one and a major contribution by Eichberg (see Brownell, 1995, pp. 17–21). However, from the perspective of an American anthropologist, the 'culture' half of the equation still seems underdeveloped: it still does not appear that Eichberg is fully using the culture concept as conceived of in cultural anthropology. That is to say, in his works the body is not as fully contextualised in relation to local symbols, beliefs, practices and history as one might expect. Hence, Nitschke's call for more contextual detail, as discussed above, is relevant to some of Eichberg's later works as well. Again, in Eichberg's defence, I would like to point out that there are bound to be lacunae in works that are experimenting with new directions. Furthermore, it is no easy task simultaneously to write a detailed ethnography and to offer a broad account of historical changes. My own book on sports in China, for example, is a detailed contemporary ethnography that outlines only a brief historical context, and it is still almost 400 pages long (Brownell, 1995).

FUTURE DIRECTIONS

One cannot predict the directions that Eichberg's future works will follow because, taken as a whole, Eichberg's opus reflects more than simple methodological or topical prescriptions: it is a way of seeing, a guide to thinking. Like the modern bourgeois order of things, Eichberg's work is temporally, spatially and directionally unbounded. So long as other scholars are creating structures, Eichberg will be exploring the places between them with the goal, as he puts it, of 'leaving power no place to hide'.[3]

NOTES

1 In American anthropology, systems of thought and belief are called 'culture'. Nitschke, however, does not seem to be working with this kind of culture concept, which is central in American anthropology.

2 I have translated *Leisten* as 'to perform' or 'performance' and *Leistung* as 'achievement' in order to differentiate the two words in English. Normally, the preferred translations are exactly the reverse: *Leisten* is 'to achieve' and *Leistung* is 'performance'. However, in Eichberg's list of idiomatic uses of *Leisten*, the English word 'performance' seems to fit better: in English one may 'perform' one's duty but one may not 'achieve' it. In choosing to translate *Leistung* as 'achievement', I am also going against a practice that has been well established since the translation of Herbert Marcuse's influential work on the *Leistungsprinzip*, usually translated as 'performance principle'. I prefer to translate it as 'achievement principle' instead. Marcuse emphasises that the *Leistungsprinzip* is that of 'an acquisitive and antagonistic society in the process of constant expansion . . .' see Marcuse, 1966, p. 45. 'Achievement' means an acquisitive urge: to achieve something is to obtain it. 'Performance' does not imply acquisition. The difficulties of translating these concepts may indicate a cultural variation between German and English speakers despite their being lumped together as 'Western civilisation' in Eichberg's and Marcuse's work.

3 I would like to thank Allen Guttmann for helpful comments on a draft version of this essay.

BIBLIOGRAPHY

Benedict, R. *Patterns of Culture* (Boston, 1989: orig. 1934)

Brownell, S. *Training the Body for China. Sports in the Moral Order of the People's Republic* (Chicago, 1995)

Carter, J. and Krüger, A. 'Introduction', in Carter, J. and Krüger, A. (eds) *Ritual and Record. Sports Records and Quantification in Pre-Modern Societies* (Westport, Conn., 1990), pp. 1–11

Dunning, E. 'Preface', in Elias, N. and Dunning, E. *Quest for Excitement. Sport and Leisure in the Civilising Process* (Oxford, 1986), pp. 1–18

Eichberg, H. *Der Weg des Sports in die industrielle Zivilisation* (Baden-Baden, 1973) (English language summary on pp. 165–72)

Eichberg, H. '"Auf Zoll und Quintlein". Sport und Quantifizierungsprozeß in der Frühen Neuzeit', *Archiv für Kulturgeschichte*, 1974, pp. 141–76

Eichberg, H. *Leistung, Spannung, Geschwindigkeit. Sport und Tanz im gesellschaft lichen Wandel des 18./19. Jahrhunderts* (Stuttgart, 1978) (English language summary on pp. 307–10)

Eichberg, H. 'Nachwort zur 2. Auflage [Afterword to second edition]', in *Der Weg des Sports in die industrielle Zivilisation* (Baden-Baden, 1979), pp. 165–75

Eichberg, H. 'Von den Exerzitien zum Sport. "Leistung" im historischen Bild', *Journal für Geschichte*, 1980, pp.20–3

Eichberg, H. *Sozialverhalten und Regionalentwicklungsplanung. Modernisierung in der indonesischen Relationsgesellschaft* (West Sumatra) (Berlin, 1981)

Eichberg, H. *Die Veränderung des Sports ist gesellschaftlich* (Münster, 1986)

Eichberg, H. *Leistungsräume. Sport als Umweltproblem* (Münster, 1988)

Eichberg, H. 'Stronger, funnier, deadlier. Track and field on the way to the ritual of the record', in Carter, J. and Krüger, A. (eds) *Ritual and Record. Sports Records and Quantification in Pre-Modern Societies* (Westport, Conn., 1990a), pp. 123–34

Eichberg, H. '"Gefährlich Denken". Über Rationalität und Angst in der Sportwissenschaft', *Stadion*, 1990b, pp. 223–55

Eichberg, H. 'Der dialogische Körper. Über einen dritten Weg der körperanthropologischen Aufmerksamkeit', in Dietrich, K. and Eichberg, H. (eds) *Körpersprache. Über Identität und Konflikt* (Butzbach, 1993), pp. 257–308

Eichberg, H. 'The enemy inside. Habitus, folk identities – and a controversial political biography', in Veijola, S., Bale, J. and Sironen, E. (eds) *Strangers in Sport. Reading Classics of Social Thought* (Jyväskylä, 1994), pp. 90–119

Elias, N. *The Civilizing Process*, Vol. 1: *The History of Manners* (New York, 1978: orig. 1939)

Elias, N. *The Civilizing Process*, Vol. 2: *Power and Civility* (New York, 1982: orig. 1939)

Elias, N. *The Court Society* (Oxford, 1983)

Elias, N. 'An essay on sport and violence', in Elias, N. and Dunning, E. *Quest for Excitement. Sport and Leisure in the Civilizing Process* (Oxford, 1986), pp. 150–74

Elias, N. *Reflections on a Life* (Cambridge, 1987)

Foucault, M. *The Order of Things: An Archaeology of the Human Sciences* (New York, 1970: orig. 1966)

Foucault, M. *The Archaeology of Knowledge and the Discourse on Language* (New York, 1972: orig. 1969)

Güldenpfennig, S. 'Gradwanderung mit Absturzgefahr. Anmerkungen zu einer linken Polemik gegen die rechte Gefahr', *Sozial- und Zeitgeschichte des Sports*, 1991, pp.63–75

Guttmann, A. *From Ritual to Record. The Nature of Modern Sports* (New York, 1978)

Mandell, R. 'The invention of the sports record', *Stadion*, 1976, pp. 250–64

Mandell, R. *Sport. A Cultural History* (New York, 1984)

Mandell, R. 'Modern criticism of sport', in Guttmann, A., Mandell, R., Riess, S., Hardy,

S. and Kyle, D. G. *Essays on Sport History and Sport Mythology* (College Station, Tex., 1990), pp. 118–38

Marcuse, H. *Eros and Civilization: A Philosophical Inquiry into Freud* (Boston, 1966)

Nitschke, A. *Historische Verhaltensforschung. Analysen gesellschaftlicher Verhaltens weisen – Ein Arbeitsbuch* (Stuttgart, 1981)

Nitschke, A. 'Historische Verhaltensforschung und eine auf Geschichtsquellen gegründete Anthropologie' in Nitschke, A. (ed.) *Rapports II. Comité International des Sciences Historiques, XVI' Congres International des Sciences Historiques* (Stuttgart, 1985), pp. 664–72

Nitschke, A. *Bewegungen in Mittelalter und Renaissance. Kämpfe, Spiele, Tänze, Zeremoniell und Umgangsformen* (Düsseldorf, 1987)

Schreiner, K. and Schnitzler, N. (eds) *Gepeinigt, Begehrt, Vergessen. Symbolik und Sozialbezug des Körpers im späten Mittelalter und in der frühen Neuzeit* (Munich, 1992)

Sjöblom, K. 'Henning Eichberg – revolutionary or National Revolutionary? Some remarks on the critics on Eichberg's theories about sport', unpublished MS

Teichmann, F. *Henning Eichberg – Nationalrevolutionäre Perspektiven in der Sportwissenschaft. Wie politisch ist die Sportwissenschaft?*, diss. (Frankfurt/Main, 1991)

zur Lippe, R. *Vom Leib zum Körper. Naturbeherrschung am Menschen in der Renaissance* (Reinbek bei Hamburg, 1988)

ESSAYS BY HENNING EICHBERG

THE BODY IN SPACE

3

THE ENCLOSURE OF THE BODY

The historical relativity of 'health', 'nature' and the environment of sport

In 1846 Adolf Spiess, who later became the leading gymnastic authority in Prussia, wrote the following description of the open-air gymnasium of the girls' high school in Basle.

> On the cheerful, leafy Peter's Square just by the school, the authorities had what used to be the sharpshooters' practice area turned into an attractive open-air gymnasium. This . . . was where the young were now to perform their physical exercises, out in the open, in natural surroundings. It would be hard to imagine a more suitable or more prettily situated exercise area. The site is well positioned for all weathers, as classes can continue in hot sunshine or rain without interruption. A spacious, open indoor gymnasium, sealed against draughts, affords both shelter and coolness and leads directly on to the open areas outside, which have been levelled and arranged in the most attractive way, and are covered in a thin layer of grass. . . . Along the side of the open-air gymnasium which butts onto the public avenues of Peter's Square, a four-foot wall of green-painted boards has been erected, as a barrier shielding the classes from outside disturbance. This is not the place to describe the different parts and different equipment of this well-provided open-air gymnasium in greater detail. . . . We will remark only that the tasteful design and slender shape of the equipment, in its dark green paint, standing in the garden-like lay-out of exercise areas and running tracks, give the whole ground a fine appearance.
>
> (Spiess, 1846)

At first glance, one might not attribute all that much significance to the green paint on the boards and equipment. A century later the same green paint reappears, however, and this time there is more to it:

> Green symbols in the green space were provided by the architects of the University of Bremen's new sports centre. The huge complex comprehends a series of fields and open spaces, a large hall, indoor

47

swimming facilities and a seven-storey 'sports tower' with rooms for administration and teaching. . . . For the halls, 400 tons of steel pipes were manufactured as pillars and roof support construction. These pipes, produced by the Mannesmann works, were decorated with green and yellow colours. Besides their static function, they form an impressive element in the visual structure of the façade, which gives an intensive optical effect. . . . The interesting profile of the swimming hall proves with its light green light construction that modern buildings nevertheless can be beautiful.

(*Rohrpost*, 1979)

The physical culture of industrial society did not only produce exercises of the body, it also necessitated the establishment of a separate environment in which to pursue them. Physical culture and building acted upon each other. This had not always been the case, nor was it by any means a matter of course that it happened.

Yet the green paint of the walls and appliances – on both the pipes and sports buildings – testified to a bad conscience, for early gymnastics and sports had originally justified themselves by being 'outdoor movements', as a 'return to nature' and 'open-air exercises'. The process of spatial separation and immurement hence ran counter to the original aims, to the educational and hygienic ideas of the founders, and to the programmes of the spokespeople for the clubs. It is the vehemence of this contradictory process that provokes the demand for socio-historical analysis. How did the space surrounding sport's moving bodies constitute itself, and what does it tell us?

OUTDOOR GAMES IN OLD EUROPE

The development of physical exercise in the Middle Ages and the early modern period had its origin in games and exercises taking place in the open air (Eichberg, 1978). From Iceland and Norway of the tenth to twelfth centuries we know of archery and spear-throwing over distances or at targets, of skiing and skating, running, swimming and wrestling. There were also ball-games, in which teams drew up against each other on the ice in battle-like formation and pitted themselves in trials of strength and wrestling contests, which could on occasions end in death for some participants. In Germany, the parish fairs in the villages and the marksmen's festivals in the towns provided opportunities for open-air games, either in the streets or in open areas in the centre or on the outskirts. Here, the pastimes and exercises included stone-throwing and snowball fights, casting flat stones at targets or over distances, stone-putting with stones weighing a hundredweight, ball-throwing contests over distances between teams – often on the paths to and from villages – hit-ball, running races, long-jump and pole vault (Schaufelberger, 1972). In the Homussen ball-game, flying objects were struck over great distances, to be caught safely by a catcher with a shingle. Wrestling, as in the form of Swiss belt wrestling, swinging or *Hosenlupfen*, was also practised in the open air. In addition, numerous pictures of peasant culture in the early modern period show dancing as a recreation enjoyed out of doors.

The equivalents in England, Wales, Scotland and Cornwall were tests of strength such as caber-tossing and hammer-throwing, but more especially violent ball-games in which whole villages formed up against each other in warlike fashion, not infrequently suffering dead and wounded as a result. The intense way in which the whole physical and social space was involved in such games is illustrated in a report by Richard Carew from 1602 of a hurling match between two Cornish villages:

> Some two or more Gentlemen doe commonly make this match, appointing that on such a holyday, they will bring to such an indifferent place, two, three, or more parishes of the East or South quarter, to hurle against so many other, of the West or North. Their goals are either those Gentlemens houses, or some townes or villages, three or four miles asunder, of which either side maketh choice after the neemesse to their dwellings. When they meet, there is neyther comparing of numbers, nor matching of men: but a silver ball is cast up, and that company, which can catch, and cary it by force, or sleight, to their place assigned, gaineth the ball and victory . . .
>
> The Hurlers take their next way over hilles, dales, hedges, ditches, yea, and thorow bushes, briers, mires, plashes and rivers whatsoever, so as you shall sometimes see 20, or 30 lie tugging together in the water, scrambling and scratching for the ball. A play (verily) both rude and rough, and yet such, as is not destitute of policies, in some sort resembling the feats of warre: for you shall have companies layd out before, on the one side, to encounter them that come with the ball, and of the other party to succor them, in the maner of a foreword. Againe, other troups lye hovering on the sides, like wings, to helpe or stop their escape: and where the ball it selfe goeth, it resembleth the joyning of the two mayne battels: the slowest footed who come lagge, supply the showe of a rere-ward: yea, there are horsemen placed also on either party (as it were in ambush) and ready to ride away with the ball, if they can catch it at advantage. But they must not so steale the palme: for gallop any one of them never so fast, yet he shall be surely met at some hedge corner, crosse-lane, bridge, or deep water, which (by casting the Countrie) they know he must needs touch at: and if his good fortune gard him not the better, hee is like to pay the price of his theft, with his owne and his horses over-throew to the ground. Sometimes, the whole company runneth with the ball, seven or eight miles out of the direct way, which they should keepe. Sometimes a foote-man getting it by

stealth, the better to scape unespied, will carry the same quite backwards, and so, at last, get to the goale by a windlace: which once
knowne to be wonne, all that side flocke thither with great jolity: and if
the same bee a Gentlemens house, they give him the ball for a *Trophee*,
and the drinking out of his Beere to boote.

(quoted in Dunning, 1971)

In the French ball-game *choule* or *soule*, church portals, walls, field boundaries
and even puddles were chosen as 'goals', around which regular water-fights
erupted, or makeshift goals or pillars were set up with a ring on them.

The spatial configuration remained the same even when it was the nobility
who were the participants, as, for example, in the Italian *gioco del calcio*, played
on the frozen river Arno in Florence and later in the city squares. *Pallone*, the
Italian game with inflated balls that was played at German courts, was an
open-air game, but towards the middle of the seventeenth century it was
ousted by court tennis when open-air activities ceased to be fashionable. And
this latter shift from the open air to an enclosed or indoors environment was
highly significant.

NOBLE EXERCISES AND ENCLOSURE IN HALLS

The spatial demarcations that characterised courtly exercise culture in the
seventeenth and eighteenth centuries had their origins in the society of the
feudal nobility. Philippe Ariès has drawn attention to the fact that from
the twelfth century the knightly jousts were probably the first social games to
which children – including those of the nobility – were forbidden entry. The
lower orders, too, were excluded, and there was hence a configurational relationship between three processes of exclusion: spatial separation, social division by rank or class, and exclusion of children (Ariès, 1976a). The embodiment of the new, enclosed space in which the aristocratic exercises of the
seventeenth and eighteenth centuries flourished was the covered hall, a space
enclosed on all sides. In the ballroom the minuet was danced; and in the
seventeenth century, the apron of the stage was inserted between the audience and the ballet as a further spatial bisection and demarcation (Eichberg,
1978). In the riding-halls, the horses rode the volte and were put through their
paces – disciplines reflected architecturally in the circles marked out on the
floor, with pillars at their centres. The fencers were accommodated in a
fencing-room where the vaulting-horse also stood. As far as the gymnasts
were concerned, who from 1810–11 were beginning to get out again into the
open air, vaulting and fencing were for a while still seen not as outdoor sports
but 'in their whole artistry' as suitable only for indoors (Bornemann, 1981).

A particularly characteristic embodiment of the process of artificialisation
and interiorisation of the body in the age of 'exercises' was the early tennis-
court, the ball-house, in which *jeu de paume*, a game not unlike tennis, was

played (Streib, 1935). From the fifteenth and sixteenth centuries onwards, these structures began to spring up, first in France then soon after in Italy, Spain, England and Germany. In the early stages of their development they appeared without roofs, but they soon became entirely enclosed. London in 1615 had fourteen indoor ball-courts; in Germany by 1600 there were seventeen such buildings, a further thirty-two being added in the seventeenth century and fourteen more in the eighteenth. A visitor to Paris in 1596 counted as many as 250 ball-courts. The architecture of the ball-court can be interpreted in various ways. At first sight one might view it simply as protection from the elements and therefore as constructed. Moreover, this aspect alone does not suffice as an explanation, since the evidence of the early form, the 'open court', suggests that the roof was not the determining and perhaps not even a primary factor in ball-court design. The architectural demarcation was, in fact, also social. By means of the court's walls, the nobility playing with their rackets set themselves apart from the people, and the walls thereby spelt out in stone what decrees in France prohibiting the lower orders (in 1452, for instance) from playing the ball-court game demanded on paper: exclusivity for one class.

But not only that, for the new architectural design was part of the very configuration of the game itself. All four walls and the roof were playing surfaces: the ball being struck against them, a great variety of possible strokes and flights of the ball resulted, creating a sort of three-dimensional billiards. Furthermore, the upper and lower walls of the playing-area were surfaces of attack and defence, containing special winning-holes or 'hazards' into which the ball could be played to score points. The spatial configuration was further emphasized by the net, the central line, the dead-ball lines and rectangular marks on the floor to make it clear exactly where the ball bounced, and by the contrast between the blackness of the walls and the white ball. From surrounding galleries the game was watched by courtly, academic or patrician spectators, and so the structure of the enclosed space was thus itself thoroughly part of the game. When chosen as the site of representative exercises, spaces outside were also organized along much the same enclosing lines. The geometrically cut proscenium stage created peephole perspectives for the ballet, while the parade-ground was the setting in which a choreography of marching bodies unfolded, just as the symmetrical castle courtyard was for the mounted quadrille. The tourney-lists were enclosed by galleries and palings, the two riders being divided by a board, the 'palia'.

A GREEN REVOLUTION AROUND 1800

Towards the end of the eighteenth century, the exercises of the nobility began to undergo a transformation and their indoor recreational culture to fall into decline. One by one the celebrated riding schools closed their doors, ending with Dresden (1848) and Hanover (1863). The only one to persist, as a kind of museum piece, was the Spanish Riding School of Vienna. Instead, people turned to outdoor activities, to formless speed-riding in the open, to fox-hunting and coursing in the English manner, to steeplechasing and horse-racing. The ball-courts also lost their attraction. Deserted, they were pulled down or transformed into hospitals, storehouses, theatres, barracks or libraries, and G. U. A. Vieth (1795) described the ball-court game as having fallen 'quite into disuse' in Germany. The rupture was finally sealed in France, on 20 June 1789, when the French Third Estate took its famous tennis-court oath, and then in Vienna in 1835 when the Viennese winter riding school was used for an industrial exhibition. Revolution and the rise of industry had soon swept aside the society to which the exercise halls belonged.

In the meantime, a quite different environment encompassing entirely new forms of physical recreation had emerged. Philanthropic teachers had taken to wandering off into the open country with their pupils, and there encouraging them to go swimming or ice-skating. Their recommending of open-air gymnasia for the schools also testified to a new pattern of body–environment relationships. Education, according to J. C. F. GutsMuths,

> thrives best in the bosom of nature. . . . Our gymnasium should, as far as possible, be the outdoors. To find a rapid, thorough cure for the weakening of mind and body that has occurred, put man straight back into his element, into the fresh air, the play of light, amongst all the influences which can work so secretly and powerfully only out-of-doors and in contrast to which all your walled-in exercise institutes are the poorest substitutes. Moreover, we want to accustom the young to the influences of the weather and what better way to do this is there than to turn them out, after their lengthy, compulsory sessions in the classroom, into the open-air exercise yard. Let us first and foremost promote public education and use that space, which has its being in itself, without doing violence to it.
>
> (GutsMuths, 1804)

To this end, in the name of healthiness and naturalness, GutsMuths recommended a grassed or sanded area in the open air, skirted by two lines of trees between which the (uneven) running-track should pass, and with various appliances scattered about, such as a tree-trunk, wooden posts and a board for balancing on, beams and a climbing pole, cross-beams and a slack rope, a long-jump ditch and high-jump poles. Where one was available, an enclosed yard or garden in the district of the school might be used, but more preferable was an area 'open to the public's gaze but within an enclosure' (GutsMuth, 1804).

Friedrich Ludwig Jahn's gymnastics started even closer to the elements and even more in the open: 'It was in the beautiful spring of 1810 that on Wednesday and Saturday afternoons when there was no school first a few pupils, and then more and more, started coming out with me into the woods and fields' (Jahn and Eiselen, 1816). It was among such woods and fields that the first gym school was provisionally established in 1811, on the 'Hasenheide', on 'true, deep Brandenburg soil', 'beneath tall spruces' and surrounded by firs and brushwood. It was equipped with a cabin and 'horizontal bars, parallel bars, vaulting and climbing equipment, with the top rope tied to a tall spruce, from which yard boys were soon attempting the riskiest manoeuvres' (Massmann, 1859). But there was no strict spatial demarcation, for 'the perimeter soon had to open up to other considerations of physical education, to running races, which were held outside the perimeter, and to games in the woods to which each swarm despatched there rushed cheering and shouting' (Massmann, 1859). For the most part emphasis was not on formalised exercises, but rather 'on games, especially "Schwarze Mann" and "Räuber und Wanderer", played between the tree-nurseries and the wooden huts' (Dürre, 1852). In 1812, Jahn started a newer, larger gym on the eastern edge of the 'Hasenheide', and here gymnastics proper assumed greater importance, even if the idea of uncontrolled play in the open air remained. The open-air gym, in Jahn's view, 'must be on solid ground covered with short grass, and be planted with trees. . . . If trees are completely lacking then some must be planted. . . .' (Jahn and Eiselen, 1816). 'No open-air gymnasium', he continued,

> should be without its playground. Also, beyond the limits of the gym-nastics area itself, each gym school should by rights have a further large area where woods alternate with open fields, where groves, bushes, shrubs, thickets and open spaces are all encountered and deciduous trees and brushwood.
>
> (Jahn and Eiselen, 1816)

Enclosed spaces were obtained for the winter, but the sense of space as not only a physical but also a social environment was still present. Thus, it was said of the indoor winter gymnastics of 1817–18 that the fencing and vaulting indoors were regarded as being in line with 'convention', and therefore more 'exclusive and less common' than in the public square (qouted in Steins, 1978). But this was the opposite of what Jahn preached and practised, since he drew an unfavourable comparison between such aristocratic forms of exercise and more 'popular' ones. A favouring of the great outdoors fitted in well with Jahn's tendency to suspect all teaching:

> It is not really possible to instil anything in people. To try to do so is like daubing wooden houses to try to make them look like marble – no durable refurbishment but just a transitory bit of decoration. What education a man [sic] can get he will obtain through his own activity.
>
> (quoted in Neuendorff, 1928)

In addition, for Jahn the liberation of space simultaneously initiated a liberation of the body, thus sparking off a reform in people's dress. Jahn and his 'long-haired louts from the Hasenheide' made the following impression on the Prussian court historian, Heinrich von Treitschke (1882):

> Their long hair hung down unkempt on to their shoulders; and their shirts were open-necked – since a servant's scarf befitted the free German as ill as an effeminate waistcoat; their broad shirt-collars covered the low stand-up collars of their dirty coats. And he complacently referred to this questionable garb as the true dress of old Germany.

Jahn's gymnastic dress of loose linen represented the radical wing of a clothing reform that was underway on a broader front, casting off the straight laces of the *ancien régime*. Similarly, his gymnastics on the green heath was only part of a wider movement out into the open that was taking place in physical recreation. Originating in England, sport in the nineteenth century broke out of the confines of the halls, taking the form of running and rowing, cycling and winter sports. From the 1790s onwards, the first German seaside resorts such as Heiligendamm (1793) arose on the English model, alongside river-based swimming pools and military swimming baths: in both cases, health considerations coincided with the outdoor movement.

GYMNASIUM AND RESTORATION

But the gymnastic 'green revolution' failed to win the day and Adolf Spiess's green boards of the 1840s were signs of yet another configuration, as buildings of a new type began to emerge around the activity of physical exercise (Heiny, 1974). The gymnasium in Frankfurt am Main, constructed in 1846 and probably the first 'Vereinsturnhalle' or club gymnasium to be built, still had twelve doors connecting it with the outside – hence making it an open hall for winter gymnastics and also allowing shelter from the rain. Soon, however, gymnastics were immured completely. From the 1850s onwards, school gyms began allocating separate parts of their building to different classes, and in the 1860s the first representative civic and 'Verein' buildings were decorated in the 'historical' style (thus recalling the old courtly ball courts).

Early accounts of the gymnastics movement generally ascribed this development to the Prussian prohibition of gymnastics, which, as one of Metternich's measures against demagogues, was enforced on gymnasts, students' organisations, democrats and national revolutionaries from 1819. But this explanation is quite inadequate. In south-west Germany, where the ban was barely effective, for instance, the same trend indoors can be observed. Moreover, were the above explanation the case, one would expect there to have been a rejection of indoor gymnastics once more after the ban was lifted in 1842, particularly given that during the period of persecution open-air gymnastics had survived as a subversive activity. Indeed, as late as 1848, revolutionary girl gymnasts were being drawn out under the 'open skies' and into the 'dark of the woods', where 'freedom stretches out around us in all the trees and branches' (*Frauen-Zeitung*, 1851). But such a return outside never took place: indeed, it was with the lifting of the ban that the building of gyms and recreation halls really began on a large scale. A glance at the history of indoor gymnasia in other countries, in Scandinavia for example, illustrates how little these buildings can be traced back to the effects of the prohibition. There too the configuration of the indoor gymnasia was part of a new pattern of relationships between the body and its environment, a pattern that had in effect been prefigured somewhat earlier – even in Jahn's own circle, by Ernst Eiselen for instance – and that involved formal exercises by rank and file, exercises on the appliances, either at timed intervals or to commands. The trees of earlier times had been replaced by gymnastic equipment and machinery, even though Jahn had in vain protested at such 'drill' as practised by Eiselen.

And this was not the only thing that pointed to factors beyond the prohibition, since simultaneously with the rise of the indoor gymnasia, there sprang up bath- and wash-houses of a new type, the forerunners of the indoor swimming pools with their divisions by sex and class – places like the 'Dianabad' in Vienna in 1842 and those in Liverpool (1842), Hamburg and Berlin (1855). The health and hygiene aspect was involved here as well, but this time it worked in the opposite direction: it forced sport and exercise back indoors. Furthermore, across Europe circus buildings were also going up in the shape of commercial enterprises such as the Cirque d'Hiver in Paris and the Hippodrome in Copenhagen (1855), the Friedrichstrasse circus (1850) and the Otto'sche Zirkus (1855) in Berlin, not to mention various dioramas, commercial sports halls and aquatic amphitheatres for water shows. Among the most famous of the sports halls were Madison Square Garden in New York, the Albert Hall in London (1867–70) and the Sportpalast in Berlin.

A SECOND GREEN WAVE

As the nineteenth century came to an end, there were signs of another reaction against all of this, a movement back 'out into the open' and once more 'back to nature'. Its most visible manifestation was the *Wandervögel* youth movement, which gave expression to its new practices in such songs as 'Wir wollen zu Land ausfahren' ('Let's go off to the country') and 'Aus grauer Städte Mauern' ('Out of the grey city walls'). Apart from the *Wandervögel*, there were the newly-formed workers' youth organisations, which were also intent on getting out into the open air. Among adults, outdoor tourism experienced a related surge during this period, whether in the bourgeois form of walking or in the proletarian mass organisation of the 'friends of nature' advocating alpinism or cycling.

Nor did gymnastics, whose practitioners were now being advised again by doctors to take to the open air, remain uninfluenced by these trends. Alongside indoor gymnastics performed on special equipment, there now emerged both a 'popular gymnastics' of track and field exercises in the open air and a games movement that adopted old games of Jahn's and newer English ones. Those who set out into the countryside on hikes, or for folk-dancing and singing, were by and large the younger members of the gym clubs who took their exercises with them 'back to nature'. As a result of a Jahn renaissance, there was much self-criticism, and one doctor hence wrote that:

> We have allowed our national form of physical recreation, gymnastics, to be driven too much out of Jahn's old open-air gym with its running tracks and slalom paths into the narrow confines of the indoor gymnasium, where long-distance running is out of the question.
>
> (quoted in Rühl, 1978)

Thus the interest of the gymnasts shifted from the closed gym, via open buildings of the type now reintroduced in Kassel, Neukölln and Spandau, to exercise areas in the open. One factor, alongside the youth and hiking movements, that contributed to this new orientation in gymnastics was a sense of rivalry with English sport. When English sport arrived in the late nineteenth century, great stress was laid on its entailing 'grass sports' and 'outdoor games' (Heineken, 1893), and as late as the 1920s it was being defined as a form of open-air movement and was contrasted as such, polemically and on health grounds, with indoor gymnastics. 'So, let us up and at it! And let no one rest until we have won back the open air for our dear German youth!' (Hessen, 1908, p. 83).

In the years after 1900, there was a parallel shift from indoor to outdoor swimming and bathing (Kusche, 1929; Wolff, 1908; Prahl and Steinecke, 1979; Heller, 1979). The sunbathing and sea swimming of the bourgeoisie, river pools and workers' swimming clubs, meant more than just a change of swimming environment or ideas of health and hygiene, because in step with these developments the image of the socially desirable body also began to alter. White skin, which had hitherto been a mark by which the nobility and others had distinguished themselves from weather-beaten farm labourers and workers, lost its value as the dominant social norm. Now it was bronzed skin and a 'sporty' appearance that exuded social prestige.

As in the days of *Turnvater* Jahn, the trend to 'open' spaces was linked to the 'liberation' of the body from restrictive clothing, and propaganda for sport as an outdoor activity did not fail to point out the liberating effect of sports clothing or the contrast with the 'indoor' character and 'restrictive clothing' of gymnastics:

> Sports wear is concerned only to be functional, although as far as detail goes the individual has considerable freedom; in this respect,

gymnastics evokes uniformity and militarism. In addition, there is the way in which the limbs are restricted by this dress, which, though light, leaves no part of the body free. The long stockings, the sleeves down to the wrist, the trousers reaching to the knee and even lower, all this adds up to a uniform.

(Risse, 1921)

Those most affected by the reform in sports dress were women, who were occasionally obliged to demonstrate their changing dress in spectacular ways, as at the 'trouser congress' of 1897 in Oxford, when female cyclists tried to get their plus-fours publicly accepted. In the sphere of the bathing costume, traditional severity and separation also began to ease, notably in the 1920s, and on stage Isadora Duncan – who first appeared in Berlin in 1902 – caused a stir with her 'free', expressive dancing, executed barefoot and in scanty dress. The most radical manifestation of these tendencies was the nudist movement, which had been burgeoning since 1900, and one of the forms in which it continued after the First World War was as a popular workers' nudist movement. The more radical and, on occasion, sect-like tendencies were only illustrations of a reformist trend in both living and clothing patterns, however, and this was one that had a broad social base and indeed reached into the ranks of the bourgeoisie. For women, sportswear played its part in offering a new alternative to the corset, while for men elements of 'liberation' came chiefly from the youth movement's ousting of the stand-up collar by the open collar and the abandonment of the bowler hat.

The extent to which this physical–spatial liberation affected other spheres of life can be seen in the open-air theatre movement that took off around 1900. In education during the same period, reformists were aiming the brunt of their criticism at the spatial configuration of the Wilhelminian school, the 'school barracks' and the tendency to model school playgrounds on parade grounds (Kraft, 1977). Walls and other enclosures, regarded as artificial and restrictive, fell from favour, as is evidenced by the open-air museums of the period or by attempts at 'garden cities' and other forms of 'natural building': the call was for a new environment, a new body, a new hygiene.

COLONISATION OF SPACE DURING THE TWENTIETH CENTURY

This second 'green wave' of the early twentieth century was also quickly arrested, and then rechannelled into other and often contrary directions. From the 1920s on, the open sports fields began to be transformed into concrete stadiums, while the indoor gym overcame its crisis and again proliferated, continuing to the present day in the form of sports and other multi-purpose recreation centres. Sport monocultures, complete with their dedicated fields and halls, moved out into the countryside. River swimming

57

was replaced by public swimming baths, a striking new example of civic building (Samel and Zepmeisel, 1928; Ortner, 1956: Fabian, 1960).

Since the 1970s, however, the ways in which sport and tourism (often linked to sport) have 'exploited' space have come in for some criticism (Krippendorf, 1975; Stern, 1976; Dieckert, 1977; Koch, 1979; Eichberg, 1982). Quite suddenly, there was a renaissance of the feeling that the space in which recreation and sport take place is of social significance, and, what is more, this space was now felt by many to be much narrower and more restrictive than it had ever been before. The practical changes that form the background to these new 'green' criticisms are highly contradictory, but three trends can be discerned.

First, a trend that has become particularly evident over the last decade has been that of immuring physical exercise, a process that has reached its zenith in windowless school sports halls, with their air-conditioning and artificial lighting, but which has also become increasingly common in the commercial sector in the form of both keep-fit studios (with their equipment parks in rooms) and squash courts (where the body communicates with the walls via the ball). One firm in Salt Lake City is already offering an indoor golf-course, a combination of slide-projection (reproducing the course at Pebble Beach in California), special cameras and microprocessors to calculate the flight of the ball. Behind such ideas one can see the further steps being taken by the video game culture of the amusement arcades. Buildings along these lines again lead to the subjection of open spaces: by motorway access roads to the sports centres, by athletics tracks made from synthetic materials, and by easy-clean 'polygrass' artificial turf for football fields. There are echoes here of the French gardens of the *ancien régime*, especially if one considers the geo-metrical concrete and steel tube constructions comprising many children's playgrounds (Gollwitzer, 1957; Ledermann and Trachsel, 1968).

Second, in contrast to such trends, the last ten years have seen new move-ments to get 'back into the open'. Running has suddenly acquired popularity, and jogging and orienteering, of American and Scandinavian origin respect-ively, have led people in different ways back out into the open air. Skiing has experienced an enormous upsurge in popularity, and two new means of exploiting space have recently appeared: the spectacular pastime of hang-gliding and the more mundane one of skateboarding. Adventure holidays and neighbourhood street festivals also illustrate the tendency towards open-air activities, but they too reveal its paradoxical character. On the one hand, the configuration of the 'back to nature' movement betrays – as in both 1789 and 1900 – connections with social protest, as seen in the open-air festivals of the young or in their alternative rural communes, with the suggestion being that new and subversive physical experiences await discovery in the great outdoors (Kükelhaus and Lippe, 1982). On the other hand, besides its protest function there is also a colonising function that – as in Biedermeier in the 1840s and the stadium-buiding phase of the 1930s – is intent on throwing up new structures. In this latter respect planners propose new tourist centres while the car industry demands more roads to satisfy leisure needs:

> To recuperate in another environment, have an adventure, make new discoveries and share in other cultures are essential activities, particularly well-suited to alleviating the pressures of everyday work, business and social spheres, and to allow repressed needs to come into play. Hence a good network of roads is required to link high-density areas to the recreation areas close to them.
>
> (Walper, 1980)

The thinking behind the 1972 Olympic Games in Munich – known as the 'green Olympics' – including their masses of new buildings, was of precisely this kind. Here an attempt was made with the help of tree surgery and moss transplantation to create a special landscape for sport. Trees were felled so that the large hill that the planners wanted would not be made to look too small; a chain of hills was 'severed by a zigzagging concrete supporting wall'; beds of all varieties of flowers were laid out; 'the buildings were de-emphasized in an unmonumental way, made visually harmless, pushed into the ground': and thus 'a whole landscape was made green and flourishing in the shortest possible time'. Green had again come to signify green paint: 'By continuing the plants into the hall from outside, the partition executed only as a climatic barrier, is played down, with its metal parts being painted green' (Gollwitzer, 1972).

Third, besides exercises that take place within walls and those that are moving out into and colonising the countryside, another kind of exercise and relationship to space has emerged over the last decade. Since the 1960s, Europe and America have both offered fertile ground for meditational forms of exercise from East Asia that combine gymnastic and spiritual elements: yoga, t'ai chi chuan, tantra, zen, as well as various techniques of health-educational and bio-energetic body dynamics. Their broad social appeal ranges from middle-aged women, the first group to be attracted in the mid-1960s (before men or young people), to the alternative youth scene of the 1970s and to more recent courses for top management. The opportunities for instruction in these techniques also vary greatly, from evening classes in small towns to commercially run fitness centres in the cities, and from youth sects to Catholic monasteries. However unclear the social significance of these disciplines may still be, one thing becoming plain is their novel use of space: they neither colonize it by moving out into it, but nor do they require its reconstruction in the manner of the leisure centres. Their space is the space available in the complexity of daily life.

Other non-European exercise forms now growing in popularity point in the same direction: consider the Eastern martial arts such as aikido, kung fu, taekwon do, viet vo dao, or perhaps Afro-dancing and its polycentrism and on-the-spot dance configurations. It is true that there are signs of these dances and exercises also being confined to mono-functional halls, centres and discotheques, but in historical terms – and particularly if these bodily activities turn out to be more than just a passing fad – their thrust, directed against the separation of specific space earmarked for sports purposes, is something new.

THE SPACE OF SPORT IS NOT 'NATURAL'

The structural metamorphoses of sports space as charted above allow us to perceive certain interconnections between physical exercise and the environment. The body, it appears, does not stop at the surface of the skin. It reaches into the space surrounding it. Yet the manner in which it extends spatially is not fixed or predestined, but historical and thereby subject to social change. Thus there existed a configuration connection between the training of the body in Prussian gymnastics – entailing exercises and drill to command or at timed intervals – and the gymnasium of the Biedermeier era where it took place. In contrast, the games and sports movement of 1900 took as its starting point the open air, even if over a period of time this different type of physical culture, outdoor sport, was transferred indoors to become physical exercise between walls.

What is the explanation for such spatial changes in sport? Many natural and technological answers have been offered from the realm of sport itself, that it is 'natural' to take shelter from bad weather, and that 'technological pressures' have made the indoor gymnasium and its associated culture necessary. Seen historically, however, it is clear that these two explanations do not go far enough. Far from it being 'natural' to seek shelter from the elements, at certain stages of social development it has been considered natural to expose the body to the elements during physical exercise. Nor was it the technological revolution of industrialisation that produced the indoor gymnasia, since these enclosed spaces had been anticipated in the exercise halls of the *ancien régime*. On the contrary, at certain climactic moments of technological progress – around 1900, for instance – society tended to react against attempts to enclose physical culture spatially or to make it more technical.

What of the 'healthiness' of the space in which sport has taken place? Historical comparison shows that there is no uniform or general answer to this question either. There have been no invariable criteria by which the environment that is medically and hygenically most suitable for physical recreation might be identified. The health experts, the doctors, have given different answers at different times: in the eighteenth century, they recommended exercises that were practised indoors, but they were also in the forefront of the open-air movement. In the nineteenth century, they again gave their blessing to indoor gymnastics, and the building of wash-houses and swimming baths was promoted by them on hygienic grounds. But then, around 1900, doctors such as F. A. Schmidt were among the driving forces behind the games and hygiene movements for open air and 'reform of living' that finally gained acceptance for open-air exercise, swimming and (ultimately) sport as a whole. The sports doctors then returned, with sport itself, indoors. Now, the latest open-air movements are recommended on medical and health grounds, and even the supposedly 'spatially neutral' exercises of bio-energetics sometimes carry earnest recommendations to get outside from health instructors and physiotherapists. Views on what sort of space is healthy therefore turn out to be socially relative, and to follow fluctuating historical re-evaluations.

'Health' or medical expertise thus provides little in the way of a solid basis for interpreting the historical and socio-ecological process involved here. The same is true of the ideas sometimes claimed to be underlying this process: that is, the attempt to derive the practical changes of the everyday from philosophical or literary ideas – from the Enlightenment, say, or from romanticism or scientific naturalism – is also flawed. It cannot be right to pin the whole of the responsibility for the nature movement of 1800 on Jean-Jacques Rousseau, for instance, and in 1900 there was not even a Rousseau to be so credited. Spiess prudishly having the girls' PT area surrounded with boards, or the act of spatial 'equality' being created for competitive sport, are such elementary and unphilosophical changes that the tools of the historian of ideas are of little help here. Rather, such changes were indelibly social. But the social nature of this spatial configuration should not be taken too narrowly or too literally. For example, while it is appropriate to cite the ban on gymnastics from 1819 and the Prussian system of the nineteenth century in any enquiry into the genesis of the indoor gym, this is not to arrive at the whole explanation. Similarly, commercial interests of the kind that, in the shape of tourism and industry, created the built-up world of the ski pistes are also only one factor to be considered among many. What general theories might then explain or help us to interpret the spatial separation of the sporting body?

PERFORMANCE PRINCIPLE, CLASS POWER AND REVOLUTION

One possibility is to take the performance principle as our starting point. An interest in being able to compare sporting performances arguably led to the technical conditions of competition being 'equalised', and hence to the homogenisation and monofunctionalisation of the space in which performances occurred. This development then brought demarcations of various kinds in its wake:

1. The withdrawal of sport indoors, protected from the effects of the elements.

2. The division of sport and non-sport, in connection with the division between work and home, working hours and leisure hours.
3. The demarcation of different sports.
4. Demarcation by category of achievement and especially by age-group, leading to the exclusion of children from many sports situations.

This interpretation identifies one central element of industrial 'achievement societies', as traced out in their approach to performance sports, but it obviously does not account for the entire social phenomenon. The aristocratic exercises of the seventeenth and eighteenth centuries and the disciplined gymnastics of the Prussian kind created indoor exercise halls for themselves without any reference to achievements of the 'faster, higher, stronger' variety.

A further, or complementary, theory suggests itself: namely, that class rule and the separation of the social classes from each other favoured the emergence of the indoor hall (Ariès, 1976b; Foucault, 1977). The upper classes of the early modern period withdrew into the exercise halls, and paralleling this withdrawal absolutism created the house of correction, the factory and the barracks to control and to organise social space. The state began to appropriate the street, which up to that point had been multi-functional – the site of play and of physical culture – and transformed it into a space for representation and traffic. The riding school and the ball-court were, like the French garden, parts of this new social geometry. Then, correspondingly in the nineteenth century, the indoor gymnasium was structurally related to and arose contemporaneously with the prison, the lunatic asylum (Foucault, 1973; Rothman, 1971) and the school house in the context of a thoroughly spatial disciplining and functioning of social life. It was therefore no accident that movements against indoor gymnastics, in the late eighteenth and early twentieth centuries, coincided with periods of revolution. When GutsMuths – describing the new gymnastics environment – set 'the open air', 'the light' and 'the public space' against 'walls', 'restoration institutions' and 'compulsion', he was also itemising elements in the programme of social and political revolution then occurring. Political power and resistance to that power found visible reflection in both body movement and the shaping of space, and 'nature' and 'freedom' were thereby interlinked.

Ein Ruf ist erklungen	A call has sounded
Durch Berg und durch Tal	Through valley and hill
Heraus, ihr deutschen Jungen	Come out, German youths,
Zum grünen Waffensaal! (. . .)	To Nature's armoury (. . .)
Und deine Burg bricht nieder,	And your prison's in ruin
Du alter Meister Jahn.	Old Master Jahn.
Die Turner ziehen ins	The gymnasts are leaving for the
grünende Feld	green fields
Hinaus zur männlichen Lust (. . .)	Out to their manly [*sic*] delights (. . .)

Und frei, wie der Aar durch die	And as free as the eagle through
Lüfte schwebt,	the air moves,
Fliegt auf der Turner am Mast.	So flies the gymnast on the pole.

(Hauff, 1970)

And yet the connection between political power and the organisation of space still does not wholly explain this process. The parcelling out of space can also be seen in the history of the museum, which developed parallel to the exercise hall: from the courtly cabinets of curiosities in the era of the exercise halls, through the classical museum of the nineteenth century in the era of the indoor gymnasium, to the open-air museum around 1900 in the era of the hiking and games movements. This parallelism perhaps hints at the existence, behind manifest political interests, of something akin to a deeper and pervasive 'order of things' (Foucault, 1970). The spatial relegation of the dead from the late 1700s onwards, entailing the move to walled cemeteries sited outside human settlements, illustrates the fact that something else was involved (Ariès, 1976b; Boehlke, 1979). And, last but not least, there took place a withdrawal of some of life's more 'delicate' activities and functions beyond the grip of power- and class-interests, to be installed in separate, private, intimate rooms and hence giving rise to nurseries, toilets and bedrooms.

All of this suggests that physical separation in sport was part of a more comprehensive configuration determined by the body and its links into daily life. Norbert Elias's concept of a 'civilising process' points along these lines (Elias, 1969, 1977), although the connotations of a linear and satisfactory evolutionary progress are certainly hard to reconcile with the complexity and colonising violence of the process. How can we best illustrate this claim?

A DIGRESSION ABOUT THE SEPARATION OF THE INFANT

One phenomenon in particular, even though its connection with the world of sport is seemingly quite remote, makes an especially powerful statement about the configuration of separation: namely, the mother's battle with her babe-in-arms, and then the physical separating-off of the child from the mother. One guide for mothers of 1938 gave the following advice:

> The younger the child, the more peace and quiet it needs. . . . It is best for the child to be put in a room of its own, where it can be alone. . . . The whole family should make it a first principle not to consort with the baby except for a good reason. . . . The most common reason why worried mothers and grandmothers are always busying themselves with the child is its crying'. The infant's crying, we are told, has various causes, which should be remedied. Among measures to be adopted are time-drill, observance of the four-hour feeding interval, and 'a regular

daily routine for the baby, which is the basis of all sensible child care. . . . On no account will we give the baby its meals early, for the child will quickly learn to achieve this again by crying, and its regular routine will be endangered. . . . If the dummy does not work either then, dear mother, you must be hard. What you must not do is take the child from its bed to carry it, rock it, push it in its pram, hold it on your lap or even suckle it. The baby realizes incredibly quickly that it only needs to cry to attract the attention of some sympathetic soul and to become the object of such care. . . . As far as possible, the child should be left, alone, in some quiet place and only brought out again for the next feed. Often only a few such trials of strength between mother and child are necessary – and the first are the crucial ones! – for the problem to be solved.

(Haarer, 1938)

It is not just the handbook's metaphors that indicate the sporting character of the scenes described: note the references to 'will-power' and 'hardness' on the mother's part, to both a 'trial of strength' and 'struggle for power' between mother and infant, to the child's crying being 'a sporting activity', and to its first 'record performance'. The space–time configuration here links the drilling of the infant with the practices of sport, particularly time-drills, sacrifice of the present for the future and the subdivision of space. Industrial society's particular violence towards children (and mothers) hence derives not only from an abstruse myth, but also from a powerful and yet variable behavioural configuration.

Variable? The separation or shutting-out syndrome, at least, was the result of a historical change. Right up until the eve of industrial society, parents and children lived together in the same space: and this was not only true in the homes of farm labourers, but occurred even where there were plenty of rooms available. Thus a lawyer's son, born in 1764, remembered from his youth:

Very spacious dwellings which kept members of a family apart from one another were not common; and those who had them only ever made occasional use of their size. By and large parents and children gathered in one room, and so children worked and played under their parents' gaze.

(Jacobs, 1840, quoted in Hardach-Pinke, 1981)

And a Brunswick patrician's son, born in 1771, wrote:

We lived in a handsome, spacious . . . house, containing something like forty rooms which, but for two, remained quite unused, locked the whole year through. There was no lack of room, therefore, and yet my father had no study of his own . . .

(von Strombeck, 1835, quoted in Hardach-Pinke, 1981)

The change from this kind of life to the separation of children in industrial society was accompanied by, and rationalised through, a change of myth. In a book of advice published in 1792, an enlightened parson launched into a polemic against the old mythology as presented by two female relatives at a woman's child-bed:

> *Katherine*: Ah, sister! It's hard to believe how many accidents and vexations a newly-delivered woman is exposed to, especially before the new baby is christened. I by my life was never on my own in the room, for fear of what the evil one, Lord have mercy, and his accomplices might get up to.
> *Eva*: Quite so! quite so! She must mind out that she doesn't have happen to her what happened to Mistress Anna. In truth, her child was exchanged during the time no one else was in the room with her, and that's how she got the poor deaf and dumb boy she has now.
> (Zerrenner, 1792)

There was a danger, then, that subterranean powers would steal a child left alone and replace it with a changeling. This not only explained the origin of the 'handicapped' child, it also reinforced the necessity of suckling, mother and other members of the family staying close together. Such 'superstitions' became the object of scorn from the enlightened experts, the doctor and the preacher:

> You good people! . . . Your child's crying comes from your having always taken it out of its cradle when it cried and now it has got used to it and always wants it to happen. So have a close look at the child and if you see nothing to make you suppose it is feeling ill let it go on crying. It will certainly not cry itself to death and afterwards it will lie there quietly and even go to sleep without being cradled. Ah, you cannot imagine how important it is to start as early as you may with the training of small children!
> (Zerrenner, 1792)

With such advice came an alternative myth of 'training' and 'habituation', one consequence of which was physical separation, since leaving the child to cry was (and still is) only possible for those present – and in particular the mother – if they sealed themselves off from it behind a wall. The nursery duly evolved as a necessary means of spatial removal. This kind of exiled space has since been succeeded by the 'school barracks' and the kindergarten, the playground, school yard and sports hall. Supporting this physical separation, there have followed contrivances such as the bottle (to avoid breast-feeding) and the rubber dummy (to avoid mouth-to-mouth contact), the playpen and the pram (to avoid having to carry the baby). The maternity clinic institutionalised the violent separation of mother and newly born baby from the very first hours, although today we know a little more about how the physical separation of mother and child is linked to child abuse.

It was no accident that the parson of 1792 and the handbook of 1938 should have ridiculed, respectively, the 'old dears' and the 'set-in-their-ways grandmothers, aunts and nannies' who in both their advice and their practice resisted spatial separation. The experienced mother, the older woman – guardian of the totality of life's interconnections – is in both documents presented as an obstacle in the path of industrial society's more rational and efficient organisation of space. The new experts were the priest and the doctor: did they just happen to be male? They were as male as the new gymnastics and the new sport. The exclusion of children and the retirement of competitive sport indoors were indications of the same historical process: the birth of both pedagogical theory and sport in the wake of the subdivision of space (Eichberg, 1984).

BIBLIOGRAPHY

Ariès, P. *Geschichte der Kindheit* (Munich, Vienna, 1976a)

Ariès, P. *Studien zur Geschichte des Todes im Abendland* (Munich, Vienna, 1976b)

Boehlke, H.-K. (ed.) *Wie die Alten den Tod gebildet. Wandlungen der Sepulkralkultur 1750–1850* (Mainz, 1979)

Bornemann, J. *Lehrbuch der von Friedrich Ludwig Jahn unter dem Namen der Turnkunst wiedererweckten Gymnastik* (Berlin, 1814, Münster, 1981)

Dieckert, J. 'Sport und Umwelt – ein Problemaufriss', in *Lehren und Lernen 7, Kongress für Leibeserziehung 1976* (Schorndorf, 1977)

Dunning, E. (ed.) *The Sociology of Sport* (London, 1971)

Dürre, E. 'Jahn', *Der Turner*, 1852

Eichberg, H. *Leistung, Spannung, Geschwindigkeit. Sport und Tanz im gesellschaftlichen Wandel des 18./19. Jahrhunderts* (Stuttgart, 1978)

Eichberg, H. 'Von der grünen Heide zur fensterlosen Halle. Sport als Landschaftsfrass und Parzellierung des Raumes', in *Jahrbuch der Turnkunst 76. Deutscher Turner-Bund 1981/82* (Frankfurt/Main, 1981)

Eichberg, H. *Die historische Relativität der Sachen. Auf dem Weg zu einer kritischen Technikgeschichte* (Münster, 1984)

Elias, N. *Die höfische Gesellschaft* (Berlin, 1969)

Elias, N. *Über den Prozess der Zivilisation*, Vols. 1–2 (Frankfurt/Main, 1977)

Fabian, D. *Bäder, Handbuch für Bäderbau und Badewesen* (Munich, 1960)

Foucault, M. *The Order of Things* (London, 1970)

Foucault, M. *Wahnsinn und Gesellschaft* (Frankfurt/Main, 1973)

Foucault, M. *Überwachen und Strafen. Die Geburt des Gefängnisses* (Frankfurt/Main, 1977)

Frauen-Zeitung 3 (1851), pp. 187–9, repr. in Pfister, G. (ed.) *Frau und Sport* (Frankfurt/Main, 1980)

Gollwitzer, G. (ed.) *Kinderspielplätze* (Munich, 1957)

Gollwitzer, G. (ed.) *Spiel und Sport in der Stadtlandschaft. Erfahrungen und Beispiele für morgen* (Munich, 1972)

GutsMuths, J. C. F. *Gymnastik für die Jugend* (Schnepfenthal, 1804, Frankfurt/Main, 1970)

Haarer, J. *Die deutsche Mutter und ihr erstes Kind* (Berlin, 1938)

Hardach-Pinke, I. *Kinderalltag* (Frankfurt/Main, 1981)

Hauff, W. '"Hoffnung" und "Turnerlust" vor 1824', in *Sämliche Werke,* Vol. 3 (Munich, 1970)

Heineken, P. *Die beliebtesten Rasenspiele. Eine Zusammenstellung der hauptsächlichsten englischen Out-Door Games zum Zwecke ihrer Einführung in Deutschland* (Stuttgart, 1893)

Heiny, H. 'Die Turnhalle, eine Sonderform der Architektur des 19. Jahrhunderts', *Kölner Beiträge zur Sportwissenschaft,* 1974

Heller, G. *'Propre en ordre'. Habitation et vie domestique 1850–1930. L'exemple Vaudois,* diss. (Lausanne, 1979)

Hessen, R. *Der Sport* (Frankfurt/Main, 1908)

Jahn, F. and Eiselen, E. *Die Deutsche Turnkunst, zur Einrichtung der Turnplätze dargestellt* (Berlin, 1816, Fellbach, 1967)

Koch, J. 'Umorientierung im Spiel – und Sportstättenbau', *Sportpädagogik,* 1979

Kraft, P. *Der Schulhof als Ort sozialen Verhaltens* (Braunschweig, 1977)

Krippendorf, J. *Die Landschaftsfresser. Tourismus und Erholungslandschaft – Verderben oder Segen?* (Bern, Stuttgart, 1975)

Kükelhaus, H. and Lippe, R. zur *Entfaltung der Sinne* (Frankfurt/Main, 1982)

Kusche, J. *Verfall und Wiederaufbau des deutschen Badewesens,* diss. (Berlin, 1929)

Ledermann A. and Trachsel, A. *Spielplatz und Gemeinschaftszentrum* (Stuttgart, 1968)

Massmann, H. 'Die Turplätze in der Hasenheide bei Berlin', *Deutsche Turnzeitung,* 1859

Neuendorff, E. *Turnvater Jahn. Sein Leben und Werk* (Jena, 1928)

Ortner, R. *Sportbauten. Anlage, Bau, Ausstattung* (Munich, 1956)

Prahl, H.-W. and Steinecke, A. *Der Millionen-Urlaub. Von der Bildungsreise zur totalen Freizeit* (Neuwied, 1979)

Risse, H. *Soziologie des Sports* (Berlin, 1921, Münster, 1979)

Rohrpost, Firmenzeitschrift Mannesmann, 1979

Rothman, D. *The Discovery of the Asylum* (Boston, Toronto, 1971)

Rühl, H. 'Das Jahnbild des "Zentralausschusses zur Förderung der Volks- und Jugendspiele in Deutschland" 1892–1921', *Stadion,* 1978

Samel, W. and Zepmeisel, G. *Gerätekunde. Einrichtung und Ausstattung von Turnhallen, Sport- und Spielplätzen* (= Handbuch der Leibesübungen 13) (Berlin, 1928)

Schaufelberger, W. *Der Wettkampf in der alten Eidgenossenschaft. Zur Kulturge schichte des Sports vom 13. bis in 18. Jahrhundert,* Vols.1–2 (Bern, 1972)

Spiess, A. 'Bericht über den Turnunterricht der Öffentlichen Töchterschule in Basel', *Allgemeine Schulzeitung,* 1846

Steins, G. *Die Berliner Hasenheide. Ihre Turnplätze von 1811 bis 1934* (Berlin, 1978)

Stern, H. 'Naturschutz gegen Menschen?', in Stern, H. *Mut zum Widerspruch* (Reinbek, 1976)

Streib, W. 'Geschichte des Ballhauses', *Leibesübungen und körperliche Erziehung,* 1935

Treitschke, H. von *Deutsche Geschichte im 19. Jahrhundert,* Vol. 2 (1882) (Leipzig, 1928)

Vieth, G. U. A. *Versuch einer Encyklopädie der Leibesübungen,* Vol. 2 (Berlin, 1795, Frankfurt/Main, 1970)

Walper, K. *Freizeit und Strasse. Aktive Freizeitgestaltung und Verkehr* (Bonn, Dusseldorf, 1980)

Wolff, C. *Öffentliche Bade- und Schwimmanstalten* (Leipzig, 1908)

Zerrenner, H. *Volksbuch* (Frankfurt/Main 1792)

4

NEW SPATIAL CONFIGURATIONS OF SPORT?

Experiences from Danish alternative planning

The human being steps straight ahead, because he [sic] has an aim. He knows where to go, he has decided on one direction, and he strides resolutely forward. The right angle is necessary and sufficient for action, because it serves to determine the space in a completely unequivocal fashion.

(Le Corbusier, 1925)

SPORT AND FUNCTIONALISM

The above statement describes exactly the visions of movement and architecture in modern sports. The streamlined body moves in a straight line in a universe of right angles. This spatial configuration is related to the self-image of sport as being a planned, controlled and regulated activity, producing results in the cognitive hierarchy of strategy, tactics and technique (Hagedorn et al., 1985; Nitsch, 1982; Weinberg, 1985). Once it is decided, the sportive way leads directly to the goal. Curved lines would be dysfunctional, a waste of space, time and energy. The ornament is the expression of a primitive premodernity, or even a crime against modernity (Loos, 1962). The sportive and productive modern human being shows a straight, upright body, not the crooked poise of the crook (Hoberman, 1989). How does the world of sports buildings relate to this configuration?

In the 1920s, when Le Corbusier and Loos expressed the configuration of modernity, modern sports grounds also exhibited their patterns. 'The most appropriate shape for a gymnastic field is the rectangle', a sports handbook stated. 'In pools and swimming-halls, where space must be saved, the rectangular form of pool is most suitable. . . . Gym hall: experience has shown that the rectangular form is the most appropriate for practising bodily exercises. . . . Play grounds should be rectangular because this affords the best utilisation of space' (Gasch, 1928). The logic expressed in these norms is clearly that of rationality, and it continues to be claimed as valid today. Sport and functionalism were, and are, going hand in hand. The point to doubt, however, could be the 'experience'. What does this mean: 'experience has shown. . .'? We can be almost certain that no laboratory experiments had compared sports in different environments. Neither is it likely that any other systematic (or unsystematic) research has led to the functionalist conclusion of the right angle. In this respect, there was (and is) no rationality in the concept of sports planning. But the word 'experience' is not misleading. It must be taken not in its scientific–experimental but in its sociological meaning, as collective practice. The modernists' statements about the human being, its straight lines and right angles, were in fact based on a societal practice. They formed the rationalising superstructure vaulting above a base of real sports activity and building practice. Society acted practically, and then produced a mythological 'rationality' to legitimate (and enforce) the process of human streamlining and rectangularity. Important work in social history could be undertaken by reconstructing this route from social body and power practice towards knowledge, following the hints of Michel Foucault (1975). A long-term history of social ecology could thus be written (Eichberg, 1982, 1988, 1990; Lyngsgaard, 1990), integrating the perspectives now appearing in a new human (and humanistic) geography of sport (Bale, 1993).

In this paper we shall keep to recent changes in the field of social ecology, because the social 'experience' of sports space since the 1970s is no longer as clear as the tendencies had been before. Danish architectural creativity has planned for sports in opposition to the functionalist container. Such alternatives have included the labyrinth and snail's shell, the oval, the ellipse, circle and half-circle, the mussel and amoeba, the ship-shaped longhouse and the magical house of the sorceress. The Danish Ministry of Culture sponsored an architectural competition for the creation of a new type of 'movement house' at Gerlev *folkehøjskole* (Folk Academy) in Zealand. It was finally erected in 1988, and is discussed further below. The search for, and realisation of, alternative structures in Denmark raise questions about the societal background to the architectural process. If the straight lines and right angles of sports architecture had been telling the story of modernity, what do the new configurations signify?

NEW MOVEMENT CULTURE AND THE SPORTS CONTAINER

What was the background to the planning of an alternative sports architecture in Denmark during the 1980s? Five sources of experience and critique can be identified: a new practice of movement culture; the critique of sports architecture; more general new tendencies in expressive architecture; specific Nordic traditions; and the discussion of feminist alternatives in building. The immediate impulse for reconsidering the conventional forms of sports architecture has come from experiences with new forms of movement. The Gerlev *folkehøjskole* had, since the 1970s, changed and enlarged its programme in correspondence with the new needs and wishes arising from a younger generation of students (Korsgaard, 1982). The same happened in other folk academies such as at Køng (Funen) and to some degree also in sports education in the state sector – for example, at the Institute of Sports at the University of Odense and at the State Institute of Physical Education in Copenhagen.

The following can be recognised as the main fields of experimental activity in the 1970s and 1980s:

1. Gymnastics arising from the so-called popular (*folkelig*) gymnastics of the Swedish Ling type in the nineteenth century have changed drastically during the twentieth century, and have recently developed into an aesthetic activity following the rhythms of jazz and rock music, with impulses from dance and pantomime (Korsgaard, 1982).
2. Expressive activities combined sport with body theatre, as stimulated especially by the Odin theatre (Barba, 1986).
3. Juggling, acrobatics and the art of the clown have entered sport from both the circus tradition and the popular culture of laughter.
4. Ecstatic and sensual practices have become attractive, notably as part of a wider crossing of borders in bodily experience, and peak and flow experiences.
5. Meditative experiences have turned attention towards the inner areas of the body through yoga, concentration and relaxation.
6. New games and old folk games have been reconstructed from village traditions, and have became especially popular. In Gerlev these are practised in the framework of a living museum or 'sport historical workshop'.
7. Oriental fighting arts like pencak silat, t'ai chi chuan, judo, aikido and kendo have introduced new means of body control and body contact.
8. Green activities with kayak, bicycle, walking, running, skiing and open-air life have also become popular.

Among these activities, only gymnastics have a tradition of taking place in the rectangular space of the traditional gymnastic hall. The others – even the new aesthetic gymnastics – are rather alienated in the square structures of the gymnasium and sports hall. While the green activities have moved out into nature, there is nonetheless felt to be a lack of appropriate built space for the various new sports listed above. At the Gerlev Idrætshøjskole the traditional running track, in contrast, now lies rather unused.

During the 1970s a wave of criticism arose against the functionalism of conventional sports architecture. This started in both Denmark and Germany at about the same time. The emphases were different in the two countries, however, and corresponded to their different national backgrounds: more social in Denmark, more ecological in Germany. In Denmark, the first critical attention was directed against the functionalism of sports architecture as a hindrance to social communication. 'Too many sports facilities have been built during the last years without taking into account the needs of community and social gathering beyond the proper training', as was claimed in a statement of a commission set up by the Danish Minister of Culture in 1970 to redefine national sports politics. And the conclusion here noted that facilities for social life, club festivities, amateur theatre, discussions and the like 'are a decisive precondition' to realise the social and cultural values of sports.

At the same time, an ecological sensitivity appeared in the ranks of the German gymnastic movement (*Turnen*). In 1974 a working committee in a congress on gymnastics and leisure demanded a critical ecology of sports. The 'outward ecology' should work for more and greener leisure parks and support local environmental grassroots initiatives, even if risking 'conflicts with selfish town-planning and capital interests'. By 'inward ecology', meanwhile, sport should critically examine its own facilities: 'Are they favourable for leisure, close to nature, beneficial for creative activities, open to outsiders, useful for games, variable and multifunctional?' (DTB, 1974).

In the following years, these first critical impulses have been supplemented by more detailed debates and studies both in Germany (Eichberg, 1981; Mützelburg and Eichberg, 1984) and in Denmark (Riiskjær 1985; Riiskjær *et al.*, 1986; Vogensen, 1982). These new critical voices have not been entirely independent of the innovative body experiences described above, but have remained rather autonomous in their logics. The essential points of the new criticism in the German debate are: the unnatural building materials of sport; separation from natural light and climate; mono-functional construction; lack of formal variety; a gigantism without relation to human proportions; and a lack of integration into existing urban space or landscape. Sport is seen to be dominated by the dictates of function, achievement and bureaucratic standardisation. The lack of cultural originality and of artistic form is compensated for by isolated 'art ornaments' on one or another wall, but mostly these are left in the shadows of the 'art' of advertisements. Turned into positive visions, the critique has now led to demands such as these:

1. Space for sport should be amenable to being used in multi-functional ways, by a variety of sports and by other cultural activities.
2. It should be changeable, following the changing needs of movement culture.
3. Its forms should be varied and use the fantasy potentials of spatial creativity.
4. Organic materials and methods of building should be preferred.
5. The inner and the outer space should overlap, so that shifts are possible between 'intimacy' on the one hand and 'freedom' (in relation to the open landscape) on the other.
6. Light and climate should be regulated in a more dynamic way.

7. The sports facilities should be related to their wider environment, integrated into urban architecture and the surrounding landscape.

The Danish discussion produced a catalogue of related demands, such as nearness and decentralisation, fortuitousness and variety, non-specialisation, mixing of activities and self-organisation.

These visions of sports architecture therefore reflected broader aspects of social reorientation: grassroots democracy in sport and an ecological critique of sports in a wider sense (Erz *et al.*, 1985; Mützelburg and Eichberg, 1984). Sport was no longer regarded as 'innocent'. The contradiction between the traditional standardising of achievement sport facilities and the new openness to a more experience-oriented body culture even became introduced into the official Danish materials on 'Planning Leisure Facilities in Communities'. There are changes going on in sport, they stated, and these are demanding innovative forms and a new democracy in sports planning.

FROM FUNCTIONALISM TO A NEW EXPRESSIVITY

The critique of functionalist space in sports has been related to concrete experiences with sport, but it is no accident that it has arisen both simultaneously with and corresponding to a general criticism of architectural functionalism (Keller, 1973; Siedler, 1964). This new wave of architectural sensibility is evidently leading to rather different programmes, some in conflict with others, but all of them are in some respects relevant to the architecture of sports:

1. An 'organic structuralism' has tried to find – against the isolation of separate 'functions' – a new 'holistic' combination, stimulated by organic structures in biological–ecological life, by philosophical structuralism and by the theory of systems (Broberg, 1979).
2. 'Natural architecture' has tried to learn from biology, from natural proportions (Maaløe, 1976) and from membranes, bone structures and other bionic elements (Frei *et al.*, 1984).
3. 'Green' or 'ecological architecture' has tried to find a new synthesis between the realm of life (energy for example) and that of built structure (Kükelhaus, 1976).
4. This could also mean that green grass, trees and other plants will be directly integrated as elements of building (Le Roy, 1978; Minke and Witter, 1983).
5. Or it could prompt giving the resident and user quite another and active role in the process of creation: 'build it yourself', or a residents' democracy (Alpass, 1985; Ussing and Hoff, 1977).
6. The starting point in alternative planning could also be placed in both the outer and near environment, in life between the houses as well as on the road (Gehl, 1980).

7. Last but not least, this could bring a new evaluation of the historical distinctiveness of regions and regional building forms, giving a kind of architectural regionalism.

The spectrum of non-functionalist approaches is broad and in some regards contradictory, but binding the different visions together is a general sensitivity to the expressive elements in architecture. The straight line is no longer accepted as the one given 'rationality', and this notion became expressed in some new building initiatives arousing international attention during the 1970s and 1980s. In Hungary, Imre Makovecz and his group of young architects began to build a sort of 'wild' architecture – circular or ship-shaped, with swinging walls of wood, mystical halls with manifold vaults – creating a biomorphic architectural language full of surprises and reminiscences. Trees can be found growing 'inside', and the curved walls seem to breathe. The buildings – often houses of culture and some of them also set aside for the use of dance and sport (as the gymnastic hall in Visegrád) – were often created by the local population as do-it-yourself projects, contrasting with the approach of the Stalinist mainstream. The stimulus for the Makovecz approach came from old Hungarian wooden farmhouse architecture, from Hungarian national romanticism around 1900 and from the ideas of Rudolf Steiner. And in Denmark, Makovecz's influence was clearly present in the planning of the Gerlev movement house in the late 1980s (Lyngsgaard, 1990; Makovecz et al., 1981).

At the same time in Yugoslavia, Bogdan Bogdanovic from Belgrade built monuments that synthesised abstract forms with elements from popular traditions. The results clearly contrasted with the still-dominant worship of the straight line, itself supposedly reflecting 'progress' and panopticism, the representative new baroque and the control of perspective in state communist monumentalism. Instead, one saw curved spaces embracing the play of Turkish Islamic ornaments, reminiscences from the farmers' architecture of the Balkans, and fictional steeples towering up from the underground. In Vienna, Friedensreich Hundertwasser got a chance to realise what soon became a focus for international pilgrimages of architects. The 'Hundertwasser House' was not a building of straight lines, no two windows being the same and right angles being avoided, and ornaments, colours and ceramics were more than just alibis on the wall. The house in itself became an ornament, vaulted by Balkan onion cupolas and living trees on the roof. For many years, Hundertwasser had argued against the straight line, in favour of a human right to form and reform one's own living space, to produce one's own *Kitsch*. Having been marginalised for a long time, the moment of societal sensibility had seemingly now arrived (Hundertwasser, 1984, 1985).

The initiatives of the 1970s and 1980s were not completely new, and the question remains whether they will – like their forerunners – stay at the margins of architectural history. However, aesthetic asides can be of sociological significance. Historically the main referents of the new curved lines were tendencies from about 1900 to the 1920s, such as architectural *Jugendstil* (Art Nouveau) and expressionism. At the same time, Antoni Gaudí experimented in Catalonia with crooked columns and oblique space (Solá-Morales, 1983), Bernhard Hoetger built his magic house in Worpswede (Golücke, 1982), and Rudolf Steiner created the wooden cupolas of the Goetheanum in Dornach, which studiously avoided the right angle (Schuyt, 1980). All of these historical asides have recently generated renewed interest, and the Steiner approach was also directly present in some of the Gerlev planning.

From a cultural relativist perspective, the recent innovations in architecture also include a socio-political symbolism connected with questions of cultural identity. The new expressivity of the curved lines, concentrated in the architecture of Makovecz (Hungary), Bogdanovic (Yugoslavia) and Hundertwasser (Austria), represents a Central Europe that has been long forgotten but is today being rediscovered. This is a cultural area that for some time had managed to withdraw from the direct impact of Western and Eastern empires and their streamlining. It remained 'Balkanised', bi- and multi-lingual and lying in national 'disorder', with Catholic churches and synagogues standing side by side and meeting Christian Orthodoxy as well as Islam. Seen from this perspective, the crooked walls of space tell a political story, distanced from the straight lines and the panoptical perspective of modernist imperial power (Foucault, 1975). For the Danish historical consciousness this means in particular a marked distance from both the Roman German Empire and Prussian state symbolism.

CURVED SPACES IN THE NORTH

Attention to such aspects of cultural identity in architecture forms the background for Nordic and historical orientations in Danish body culture. Which aspects of Nordic or Danish heritage can find a new momentum under actual (postmodern?) conditions today (Lyngsgaard, 1990; Korsgaard, 1982)? The fact that the influence of Makovecz at the international level seems to be strongest in Finland maybe turns our attention in this direction. The most famous name of the 'organic' school in the north had been Alvar Aalto. His ideas came from functionalism and never left it completely, forming an original Finnish compromise, visible, for example, in the asymmetrical sports hall in Otaniemi (Lyngsgaard, 1990). The specific version of Danish functionalism, with its rather 'soft' design form and the use of wood and brick (as is evident in many sports halls such as that at Gerlev, 1962), represented another compromise (Faber, 1978). In general, such compromises were far from the ethos of Western European 'brutalism' (Banham, 1986).

Taking a further step back in history, one meets national Romanticism around 1900. One of Denmark's most remarkable buildings was the gymnastic hall of the Vallekilde Folkehøjskole, created in 1884 from the plans of Martin Nyrop, the architect of the Copenhagen town hall. The Vallekilde hall – built in wood and ornamented with pictures from Nordic mythology – has been the place where the breakthrough of *folkelig* (popular) gymnastics in Denmark has occurred. Restored in the 1980s, the Vallekilde hall has been a source of stimulation for the alternative planning effected in Gerlev. In other ways, non-traditional (though 'traditionalist') forms, mythological ornamentation and timber architecture have been combined in the bizarre Norwegian 'Dragon style', in the Swedish Darlar style and in Finnish East Karelian romanticism. The gymnastic hall of Vallekilde takes us back to the assembly halls of the Danish farmers, erected all over the country from 1870 and used for gymnastics, *folkelig* political assemblies and a wide variety of social activities. Some of these halls were octagonal, and can today be regarded as a contrast with rectangular container architecture.

One further step back one finds medieval round churches in Bornholm and elsewhere in Denmark. The Finnish vaulted wooden churches (reminiscent of Makovecz's magic houses) and the Norwegian stave churches represent another contrasting language of space. The same is also true, in a different way, for the circle around Viking fortifications of the Trelleborg type (Cohen, 1965). A question generated in discussion – yet without an unequivocal answer – is why the large longhouses in these (sacred) fortresses had no straight walls, but offered a curved shape like huge ships. Which rationality was expressed by this spatial pattern? 'Trelleborg', the name given to this type of round fortress with their curved longhouses, means 'labyrinth'. The labyrinths were curved stone settings forming a winding route from outside towards a central point inside, and in the Nordic countries several hundreds of labyrinths have been identified and reconstructed, some of them dating back to the early Iron Age. The use of the ancient labyrinth is by no means clear, but it seems to have been a place of ritual dance. As running events and ball-play are also documented, the labyrinth could possibly be called the earliest Nordic 'sports ground' (Eichberg, 1989; Kern 1982). Labyrinths have recently attracted renewed interest in Denmark (and in other countries), and new labyrinth stone settings have been erected and are used for dance and play at the sports academy in Gerlev by a local sports club in Copenhagen, and by schools and private institutions in different parts of the country. This development was prefigured by the Danish painter and situationist Asger Jorn, who made a large labyrinth collection and documented the significance of this pattern in relation to the *avant-garde* or what today might be called the postmodern. And with the labyrinth, we can return from the architectural past to processes of contemporary change. In 1990, the sculptor Niels Guttormsen erected a communal house for a Danish co-operative in Egebjerg, giving it the form of a bird's wing, made entirely of wood. This 'wing house' is related to Nordic traditions of shipbuilding, and it contrasts sharply with the modern principle of 'montage building without identity' (Dam, 1990).

THE GENDER CULTURAL RATIONALITY OF THE SORCERESS

The gender question had, from the start, a strong impact on the Danish debate about sport and architecture in the 1980s. The feminist 'red stocking' movement from the 1970s had sharpened the consciousness of gender relativity in all fields of social life. So it could not be overlooked that the movement activities that were found to be most out of place in the rectangular universe of sports halls were those preferred by women: 'health pedagogics' and yoga, aesthetic gymnastics and music, body theatre and expressive sports. Could this be purely accidental?

Again, a historical experience helps to elucidate the problem. During the 1920s to the 1940s, Danish gymnastics had been characterised by a tension between two schools, Ollerup and Snoghøj. The *folkehøjskole* in Ollerup, founded by Niels Bukh in 1920, trained students in vigorous exercises of power, discipline and agility, and required them to move in rank and file under the command of their leader, who stood high on a podium. The Ollerup architecture was correspondingly straight-lined, axial and monumental, neo-classical or neo-baroque. The Snoghøj *folkehøjskole*, founded by Jørgine Abildgaard and Anne Krogh in 1925, was characterised, on the other hand, by a more Nordic romantic architecture and avoided all monumental classicism. Their gymnastics were relaxed and swinging, accompanied by music (which was strictly rejected by Niels Bukh). Its configuration was often expressed, when outdoor space was used, in circles and winding chains. Ollerup gymnastics was very masculine in appearance, although women's teams also took part, whereas the Snoghøj school was exclusively for women.

The body cultural differences hinted at here also had a political dimension. Ollerup gymnastics and Niels Bukh, though no National Socialist himself, were orientated towards Nazi German pedagogics (and *vice versa*), whilst Snoghøj gymnastics represented a Nordic democratic vision of *folkelig* (popular) freedom, and became an ideological centre of resistance under the German occupation of 1940–5 (Korsgaard, 1982).

This historical example illustrates a gender relativity bound up with both movement culture and architectural space. Gender difference – 'female' and 'male' – is here understood not as a 'biological' fact, but as a sociological differentiation between different life forms in industrial society. Sports as well as architecture are not neutral in relation to gender. But what does this mean for an alternative, 'female' architecture, one contesting the hegemony of 'male' building patterns? This question has developed in Danish feminist *milieux* and in other Nordic countries (Birch, 1984), just as it has been discussed in Germany (Erlemann, 1973; Kennedy, 1981). Some of their theses, based on phenomenological reflection and ethnological observations from matrilineal societies, have run as follows:

1. Female architecture is more holistic and complex than that of the male, given the latter's tendency towards specialisation and one-dimensional mono-functionalism.
2. Female architecture follows the principle of change and variability, rather than a definitive determination of form and function.
3. Female architecture is organic rather than systematically constructed in an abstract (male) manner.
4. Female building is more determined by use and application than by the outward effects of (male) representation. It works from the inner to the outside, instead of the inverse.
5. Female space is developed by the user, rather than by the designer, and has hence had a tendency to be anonymous.
6. And this is why female buildings grow comparatively slowly, rather than from the rapid result of quick decision and construction.

The alternative principles underlying female space produced round and curved forms – circle, ellipse, oval, spiral, half-circle, U-form, labyrinth – rather than the right angle and the straight line of male architecture. Circle and arrow symbolise the two different spatial orientations. Where female architecture uses the circular form, this is not derived from abstract geometrical logic as is the panopticon, the citadel or the prison, with their central perspective and the 'view of power' (Foucault, 1975); instead, the female architectural circle is set more in a family with the uncalculable curved line of 'life', of 'chaos' and of the Nordic 'sorceress magic'. All this is not meant to confront the rationality of (male) functionalism by a sort of (female) 'irrationalism', but to reconstruct another rationality altogether (Bloch, 1985; Irigaray, 1977). To establish a plurality of rationalities is also the authentic contribution of this debate to the spatial critique and development of sports.

The archetypal figure of the *hekse* (sorceress, witch, female shaman) – which in Denmark does not carry the possible negative connotations found in the English language – has gained a new currency from the recent feminist literature (Drefke and Fritsch, 1981), and has thereby given its name to the first project at Gerlev Folkehøjskole. In 1983, an outline was published under the working title '*hekse* house' (also in Norwegian, Finnish and German) appealing especially to feminist architects in Denmark (Eichberg, 1983). As a professional field, feminist architecture – represented by the Thyra architectural bureau in Copenhagen – was (and still is) rather weak in Denmark, but the result of the Gerlev architecture competition showed that women contributed five of the thirteen most innovative projects. This is rather remarkable considering the strong male domination still existing in Danish architecture in general (with nearly 80 per cent of the members of the Danish Architects' Association being men: see Bay *et al.*, 1971).

77

AN ARCHITECTURE COMPETITION AND THE 'MOVEMENT HOUSE'

On the basis of the above experiences and reflections, the Danish Ministry of Culture decided in 1986 to sponsor an architecture competition for a new movement house and the erection of such a building in Gerlev. Some twenty-six projects were entered, and the winning design was finally built in 1988, measuring 22 metres in diameter and costing Kr.6.8 million to build. The ministers of culture and of education took part in the opening festivities. The thirteen most interesting projects were published, together with the jury's evaluations and other commentaries (Lyngsgaard, 1990; Riiskjær and Eichberg, 1989). The results showed the following profile:

1. The functionalist 'modern rectangle' of sports, albeit not really encouraged by the announcement, was present in very few projects.
2. Neo-classicist variants of the rectangle as they had traditionally been related to the gymnastic hall were absent, but some of the projects showed geometrical–classical traits.
3. Of experimental character were some 'postmodern' projects, where the straight line was broken up and arranged in a new order: as deconstructivist poetical chaos (second prize); as a Steiner-influenced space sequence; as a New Age series of symbolic spaces.
4. Quite another solution was tried by several experimental projects, which formed non-traditional spaces by deploying curved and irregular shapes: oval and ellipse, snail's shell, mussel and organic–amoebic forms. This solution corresponded to ideas about a '*hekse* house', being rather daring and eccentric but not yet mature enough for constructive realisation.
5. A third solution was tried by projects with a more harmonic character, by using cubic space, circle forms or longhouse shapes to produce the impression of balance and Danish regionalism, creating a sort of 'neo-traditionalism' turning against 'traditional' functionalism. The jury gave the first prize to one of these projects, and this winning entry attempted a synthesis between circle, round movement space, cubic house structure and pyramidal roof, perhaps a synthesis between 'male' and 'female'. Some observers claimed this to be a typical Danish compromise.

The use of the new movement house since 1988 has fulfilled or even exceeded the original expectations. With the exception of the strictly formalised ball sports and competitive athletics, nearly all indoor activities prefer the new movement house. The space appears very multi-functional in character, changing with different activities and variations in lighting and sound arrangements: it can be a cathedral of spiritual gymnastics, an arena of Turkish oil wrestling with shrill Janichar music, a crypt with darkened windows, a dancing room with open views over the landscape down to the shore of the Great Belt, a place for jazz concerts, for *avant-garde* theatre or for evening use as a café. Visitors frequently come to Gerlev because of the movement house, and other schools have expressed the wish to build in a similar 'non-traditional' way. The problematic economic situation in Denmark, however, hampers putting most of these ideas into practice.

THE 'NEUE UNÜBERSICHTLICHKEIT'

The differentiated profile of Danish experimentation – deconstructivist break, *hekse* eccentricity and anti-traditional traditionalism – makes the crisis of functionalist sports architecture highly visible. It is an indication of contradiction and change, but its relevance remains controversial. How can a case study be related to more general social processes?

An unequivocal solution to this problem does not exist. Conservative positivism repeats again and again the illusion that quantitative methods would bring a solution or avoid 'over-interpretation' (Digel, 1991). So it should be stated that the case study of Gerlev concerns just one among approximately 4,000 sports and gymnastic halls in Denmark, or 0.0003 per cent of these halls (Hansen and Povlsen, 1989), but it does amount to 100 per cent of the Ministry-sponsored architecture competitions for Danish sports buildings. Both quantifications, like any others, are relative. The question of marginality or significance cannot be decided in this way. Alternatively, the architect's competition in Gerlev could be compared with another one, perhaps the international competition for a new Danish national stadium in Copenhagen. This has been won by a characteristic functionalist announcement, including highway-related traffic planning, but this comparison would not be 'objective' either, nor even just 'subjective', but relative. This relativity requires more modesty than planning-oriented positivism would concede.

Another mode of analysis could be by theoretical (de-)construction. This is the comparison between configurations, related to the comparative phenomenology of Michel Foucault. The straight, broken and curved spaces in sports tell the history of power, of discipline and subversion, and they can be related to the sociological impressions formulated by Jürgen Habermas (1985) when explaining his alarm about what he calls the 'Neue Unübersichtlichkeit' of the postmodern situation. The 'new impossibility of survey', the new non-panoptical situation, would be confirmed (or contested) from the base of social practice, not from social–philosophical reflection alone but from the real practices of sport and building. This discourse could then contribute to the open question of whether the 1980s and 1990s experience comprises modernity or postmodernity in sports, and, of course, in society more broadly.

The discussion about modernity and postmodernity normally proceeds in rather dualistic terms, whether in concepts of valuation (good versus bad) or of evolution (old versus new). The present reflections could also be (mis)understood in this way: functionalist versus alternative planning, the straight line versus the curved line, 'male' versus 'female'. This would be as reductionist as dualistic concepts normally are. That is why some Danish sociologists prefer to employ a 'trialectic' model, describing the field of sports as a contradiction between no less than three configurations: sport as (a) production of results and records; (b) social and hygienic disciplining; and (c) dialogical body language. On the level of space, one finds correspondingly: (a) the space of achievement; (b) the psycho-hygienic space; and (c) the space of experience. These three dimensions do not exclude each other, but they constitute a societal field of tensions, one marked by hegemony and conflict. The normative dominance of achievement sport and the container shape of its space are hence intimately related to one other, as is visible in the classical modern sports hall. With the Danish gymnastic hall, however, other and conflicting traditions coexist. The planning of an experience-oriented space of movement is, from this perspective, less an 'alternative' and more the search for a third way (a subject returned to in Chapter 7).

A further analytic path could lead into the field of the social psychology of space. An experimental project into space experience and sport pedagogics at the University of Odense has been promising (Borghäll *et al.*, 1988). Danish psychoanalytic deep hermeneutics – with roots in the concepts of Norbert Elias and Alfred Lorenzer – are leading to further methods of analysing the relations between subject, society, movement and space (Nagbøl, 1986). The sociological question of sport and architecture here turns out to be a question of a (transmodern) anthropology: how do we see the image of hu(man)ity?

ALTERNATIVE AND COMMERCIAL CONFIGURATIONS

Developing sociological questions on the base of case studies and comparative methods is an effective way to dissolve conservative statements about 'quantitatively' guaranteed 'systems' and 'functions'. The comparison between the functionalist cult of the straight line and an alternative search into the 'Neue Unübersichtlichkeit' is just one possibility. Another could be the comparison between alternative planning and certain commercial tendencies that are developing at the same time (Dietrich and Heinemann, 1989).

For example, until the 1970s swimming halls had been built in Denmark (as elsewhere throughout Europe) in large quantities, mostly communal, growing from 76 halls in 1971 to 225 in 1987 (Riiskjær, 1988). Though more than 90 per cent of the users did not need facilities of international sport standards for swimming, the halls were mostly constructed on this monotonous pattern. In connection with the oil crisis in 1973, these halls then fell into an economic crisis, and it was found that users were deserting the one-dimensional swimming basins. At the same time, commercial enterprises developed new 'leisure pools' and 'bathing landscapes' that combined different recreational functions. The new architectures involved here used exotic fantasy, curved forms of basins and halls, and differentiated passages between outdoors and indoors: all refined techniques of the 'Neue Unüber-sichtlichkeit'. Users of these facilities were excited by new bodily experiences of hurtling, sliding, plunging and splashing. These artificial landscapes became – though rather expensive in terms of entrance fees – an economic success, and several communal swimming halls followed the new shape. After England, the Netherlands and Germany (Brinkmann, 1989; Hess, 1989), the new cultural phenomenon reached Denmark (Lyngsgaard, 1990). Several elements contained in the spatial configuration of the commercial leisure pools can be recognised from the alternative planning discussion: the break with the straight line and with the rectangular container principle, the sensibility for body experience rather than the demand for standardisation, the neglect of the 'achievement aspect', the mixture of 'functions', and also the new emphasis on fantasy and variety. All of this underlines the social significance of the process that has been described here in this Danish case study.

But the comparison can at the same time be a warning against naïve expectations. Could alternative planning, significant though it may be, have been a rather marginal part of what has been going on in the world of commercial sports? Are these constructions just a field of experimentation for the market? Or, even more problematic, have they been historical late-comers, while the commercial sector has been more sensitive at earlier times? Do they simply make 'containment' more pleasurable, offering a 'bread and circuses' form of social control? Whatever the answer may be, questions about modernity, postmodernity and transmodernity in the space of sport have been posed and remain open. The functionalist anthropology of 'modernity' in the sense of 1925 is no longer valid. And the material of sports shows that this is not only an intellectual problem of the 'postmodernists versus Le Corbusier' type. Rather, sport makes visible a pre-philosophical base of experience and practice that must be taken into account by any more sociological analysis (Dietrich and Landau, 1990).

In 1947 the Danish sports physician Ove Bøje formulated an imperative for the sportsperson: 'Don't waste your time after training by loafing about on the sports ground: train – and then go home!' This was as trivial as it was characteristic, relating to the one-dimensionality of modern sports as well as to a form of spatial disciplining. The actual tendencies in today's space of sports allow new questions: 'Isn't it just fun – and, by the way, social – to waste time before and after training? Let's loaf about! Shouldn't we even transform the sport space so that this waste could be furthered?' But this scenario would mean that the language was changing as well. 'Sports ground' would no longer be sports ground, 'training' no longer training and 'waste' no longer waste. And the change would be societal.

BIBLIOGRAPHY

Alpass, J. *Beboerstyret miljoforbedring* (Copenhagen, 1985)

Bale, J. 'Sportens rationelle landskaber', *Centring*, 1985

Bale, J. *Sports Geography* (London, 1989).

Bale, J. *Sport, Space and the City* (London, 1993)

Banham, R. *Brutalismus in der Architektur* (Stuttgart, 1986)

Barba, E. *Beyond the Floating Islands* (New York, 1986)

Bay, H. *et al. Women in Danish Architecture* (Copenhagen, 1991)

Birch, K. 'Længslen efter rum', in Hansen, N. L. *et al.* (eds) *Kvinder og idræt* (Copenhagen, 1984)

Bloch, C. 'Synet og berøringen', *Centring*, 1985

Bøje, O. 'Idrætslægens Raad til de Unge Idrætsudovere', in Buch, R. (ed.) *50 Aars Idræts-Kavalkade* (Denmark, 1947)

Borghäll, J., Eichberg, H. and Worm, D. *Idræt og miljø i sociologisk og pædagogisk belysning* (Odense, 1988)

Brinkmann, A. 'Das Schwimmbad geht baden!', in Dietrich, K. and Heinemann, K. (eds) *Der nicht-sportliche Sport* (Schorndorf, 1989)

Broberg, P. *Strukturalisme* (Copenhagen, 1979)

Cohen, S. *Viking Fortresses of the Trelleborg Type* (Copenhagen, 1965)

Dam, H., 'Han bygger en skulptur stor som et hus', *Information*, 4 August 1990

Dietrich, K. and Heinemann, K. (eds) *Der nicht-sportliche Sport* (Schorndorf, 1989)

Dietrich, K. and Landau, K. *Sportpädagogik* (Reinbek, 1990)

Digel, H. 'Review of Eichberg's *Leistungsräume* (1988)', *International Review for the Sociology of Sport*, 1991

Drefke, H. and Fritsch U. 'Tanz der Hexen', *Sportpädagogik*, 1981

DTB, *Kongressbilanz Aktive Freizeit, Ein Kongress des Deutschen Turner-Bundes* (Hamburg, 1974)

Eichberg, H. 'Von der grünen Heide zur fensterlosen Halle', *Jahrbuch der Turnkunst* 1981

Eichberg, H. 'Stopwatch, horizontal bar, gymnasium. The technologising of sports in the eighteenth and early-nineteenth centuries', *Journal of the Philosophy of Sport*, 1982

Eichberg, H. 'Die Hexenhalle. Frauenarchitektur für den Sport?' in Otto, F. *et al.* (eds) *Subjektive Standorte in Baukunst und Naturwissenschaft* (Stuttgart, 1983)

Eichberg, H. *Leistungsräume. Sport als Umweltproblem* (Münster, 1988)

Eichberg, H. 'The labyrinth. The earliest Nordic "sports ground?"' *Scandinavian Journal of Sports Sciences*, 1989

Eichberg, H. 'Race-track and labyrinth. The space of physical culture in Berlin', *Journal of Sport History*, 1990

Erlemann, C. 'Was ist feministische Architektur?' in Pusch, L. (ed.) *Feminismus* (Frankfurt/Main, 1973)

Erz, W. *et al. Sport und Naturschutz im Konflikt* (Bonn, 1985)

Faber, T. *Architektur in Dänemark* (Copenhagen, 1978)

Foucault, M. *Discipline and Punish* (Harmondsworth, 1975)

Frei, O. *et al. Subjektive Standorte in Baukunst and Naturwissenschaft* (Stuttgart, 1984)

Gasch, R. (ed.) *Handbuch des gesamten Turnwesens und der verwandten Leibesübungen* (Vienna, 1928)

Gehl, J. *Livet mellem husene* (Copenhagen, 1980)

Golücke, D. *Bernhard Hoetger* (Worpswede, 1982)

Habermas, J. *Die Neue Unübersichtlichkeit* (Frankfurt/Main, 1985)

Hagedorn, G. *et al.* (eds) *Handeln im Sport* (Clausthal-Zellerfeld, 1985)

Hansen J. and Povlsen, J. 'Sport er også andet end fodbold', in Nellemann, G. (ed.) *Dagligliv i Danmark i vor tid* (Copenhagen, 1989)

Hess, I. 'Bade-Landschaften', in Eichberg, H. and Hansen, J. (eds) *Körperkulturen und Identität* (Münster, 1989)

Hoberman, J. 'Sport und Körper in der Romanwelt Aharon Appelfelds', in Fischer, N. (ed.) *Heldenmythen und Körperqualen* (Clausthal-Zellerfeld, 1989)

Hundertwasser, F. *Schöne Wege. Gedanken über Kunst und Leben* (Munich, 1984)

Hundertwasser, F. *Das Hundertwasser-Haus* (Vienna, 1985)

Irigary, L. *Ce sexe qui n'en est pas un* (Paris, 1977)

Keller, R. *Bauen als Umweltzerstörung* (Zürich, 1973)

Kennedy, M. 'Natürlich Bauen mit Frauen', in Frei, O. *et al.* (eds) *Natürlich Bauen* (Stuttgart, 1981)

Kern, H. *Labyrinthe* (Munich, 1982)

Korsgaard, O. *Kampen om kroppen* (Copenhagen, 1982)

Kükelhaus, H. *Unmenschliche Architektur* (Cologne, 1976)

Le Roy, L. *Natur ausschalten – Natur einschalten* (Stuttgart, 1978)

Loos, A. *Sämliche Schriften* (Vienna, 1962)

Lyngsgaard, H. *Idrættens rum* (Copenhagen, 1990)

Maaløe, E. *Towards a Theory of Natural Architecture* (Copenhagen, 1976)

Makovecz, I. *et al. Tradition and Metaphor. A New Wave in Hungarian Architecture* (Jyväskylä, 1981)

Minke G. and Witter, G. *Häuser im grünen Pelz* (Frankfurt, 1983)

Møller, J. *Gamle idrætslege i Danmark* (Gerlev, 1990)

Mützelburg, D. and Eichberg, H. (eds) *Sport, Bewegung und Ökologie* (Bremen, 1984)

Nagbøl, S. 'Macht und Architektur', in Lorenzer, A. (ed.) *Kultur-Analysen* (Frankfurt/Main, 1986)

Nitsch, J. 'Handlungspsychologische Ansätze im Sport' in Thomas, A. (ed.) *Sportpsychologie* (Munich, 1982)

Riiskjær, S. (ed.) 'Arkitektur – Krop og sport', *Centring*, 1985

Riiskjær, S. *Kommunerne og fritiden* (Copenhagen, 1988)

Riiskjær, S., Bøje, H. and Hasløv, D. *Idræt – kulturpolitik og planlægning* (Copenhagen, 1986)

Riiskjær, S. and Eichberg, H. (eds) *Bevægelse i arkitekturen* (Gerlev, 1989)

Schuyt, M. *et al. Rudolf Steiner und seine Architektur* (Cologne, 1980)

Siedler, J. *et al. Die gemordete Stadt* (Berlin, 1964)

Solá-Morales, I. *Gaudí* (Stuttgart, 1983)

Ussing S. and Hoff, C. *Huse for mennesker. Om organisk byggeri* (Copenhagen, 1977)

Vogensen, N. 'Idræthallens tekst', *Centring*, 1982

Weinberg, P. *Handlungsorientierte Bewegungsforschung* (Cologne, 1985)

BODIES, CULTURES
AND IDENTITIES

5

SPORT IN LIBYA

Physical culture as an indicator of societal contradictions

(*with Ali Yehia El Mansouri*)

> I have never been interested in politics. Not even today. I am no politician. I am a revolutionary.
>
> (Gaddafi, quoted in Hard, 1983)

This statement by Muammar al-Gaddafi is one among many that Western observers find difficult to understand. Libya – or, as it is officially called, The Socialist People's Libyan Arab Jamahiriya – continues to cause anxiety and yet to fascinate through its contradictions. To begin with, one finds contradictions between Libyan day-to-day reality and the stereotypes about which newspapers in the West write. One reads in them of 'Gaddafi's hell' or of a 'Terrorists' state', but meets instead a calm people immersed in Islam, wearing traditional bedouin dress, even in the cities. One expects to find a whimsical man's military dictatorship, yet meets instead a 'revolutionary thinker' officially freed of all state commitments, who at times withdraws himself to the desert to meditate; and one can from time to time approach him physically very closely without being searched for weapons.

But neither do the many attempts of the Libyan Revolution at self-portrayal resolve these contradictions. On the contrary, they create new ones. One has, it is often pointed out, done away with political representation, with government, with the domination of a party and so on, in order to allow the people to rule directly. And yet, on revolutionary committees one finds new groups of leaders being formed. It is asserted that socialism has come to stay, yet German business firms report that influential persons line their pockets with unofficial payments in return for the concessions that they gain for their projects. All signposts at Tripoli Airport are in Arabic, emphasising national pride and prestige, yet one finds in 'people's stores', which visitors are told embody the Revolution's victorious achievements, an almost exclusive representation of the international market's consumer goods: European men's suits and European wedding dresses, plastic animal toys made in Italy and Danish lego construction pieces. Can a critical social science explain such contradictions? (There is no shortage of writings on this matter: see Wright, 1969; First, 1974; Fathali and Palmer, 1980; Mattes, 1982; Ahmad, 1969.)

Recent research has given rise to the hypothesis that an analysis of sport and physical training enables one to form a balanced view of societal contradictions, of processes involved in change and patterns of behaviour in a given society (Eichberg, 1981). Can one then, on the basis of an analysis of sport in everyday life, dig beneath the ambiguity of ideological statements, of their justification or rejection, and in the light of a new 'historical materialism' reconstruct the outlines of a foreign culture?

SPORT AND THE 'GREEN BOOK'

Anyone interested in the official view taken of sports in Libya will at once be referred to the *Green Book* (Gaddafi, 1980). This document appeared in 1975 at the height of the Libyan People's Revolution, when many heated debates and discussions were going on. It contains programmatic statements claiming the birth of a 'third theory' as an alternative to both Western capitalism and East European state communism, and is divided into three parts. The First is a draft plan of democracy from below, rejecting political representation of any kind, parliamentary, governmental, or domination of a party or class, since all of these forms of representation share one thing in common: the dictatorship of a part of society over the whole. Instead, the masses themselves ought now to rule and express their will directly through people's congresses, in which everyone can take part. The Second Part enlarges on socialism as a 'solution to the economic problem', the presupposition here being the need for full economic self-determination, a kind of socialism pointing to a 'third road' independent of Western and Eastern ideologies. This kind of socialism will come into its own when wage-labour is abolished, when non-occupier ownership of land and houses is done away with, and when workers themselves have their own genuine self-administration. The Third Part provides a summary of the theory of cultural revolution, and has a list of demands that the Revolution wants to see fulfilled, such as equality of the sexes and the protection of minorities, as well as statements on religion and education.

The Third Part, which deals mainly with cultural and national aspects of everyday life, also contains a chapter on sport. This chapter includes three important propositions. First, sport, like prayer, cannot be delegated to anybody else and is an activity that everyone can participate in. Thus defined, this proposition becomes an attack on sports clubs as 'instruments of social monopoly', and, framed by its political corollary, such a proposition can also be seen as an attack on political parties. Second, sport means activity and not passively looking on: otherwise the masses would be left merely sitting

in the stands, [where they] are lazy and cheer the heroes, who have robbed them of their initiative. . . . Originally grandstands were meant

88

to serve as barriers between the spectators and the playing-field. This meant that the masses were kept back from entering the playing-field. When the marching masses engage in sport and seize the centre of the playing-fields and open-areas, stadiums will be deserted and fall into disuse. . . . The grandstands will disappear, since there will be no one there to care to take a seat.

(Gaddafi, 1980, 1983)

In this connection Gaddafi refers to the specific Libyan–Arab root position from which his judgement stems: namely, that bedouin folk were not interested in the theatre or other kinds of show, and would themselves 'rather take part themselves in games and merry making'. In another context Gaddafi intensified his criticism of sport sites even more: in 1978, alongside stands and stadiums, he also cast playgrounds and swimming pools as relics of conservative thinking and hence inherently counter-productive (Fathali and Palmer, 1980). Third, and finally, the *Green Book* comes out against boxing and wrestling, calling them brutish and barbarous. They also would disappear in time in the train of civilisation, in the same way as had already happened to human sacrifice and pistol duelling.

From such reflections, the Libyan theory of sport developed a fully con-trasting juxtaposition between a new kind of desirable public sport and the less-favoured conventional one. The latter is characterised by commercialism and professionalism, by rivalry and showmanship, by purchasing expensive equipment, social exclusiveness, and also by doping through drugs and exploiting children. Public sport, on the other hand, was intended to improve health, education and relaxation. By using simple sports equipment that has perennial value, public sport would definitely be suitable for both sexes and for different age and social groups, which would thus reflect political and economic equality demanded in the names of both democracy and socialism (Mansouri, 1982).

And yet there are other official Libyan publications where precisely the kinds of conventional sport that have been severely criticised are held up as models worthy of emulation. In self-advertising Libyan picture books, one sees stadiums (which Gaddafi himself has decried) hailed as outstanding achievements of the 'Revolutionary State': anything from football stadiums at the bottom rung of the sports ladder to the gigantic 'sport cities' in Tripoli and Benghazi (Aljamhiriya, 1979). *Med-Post* magazine, one of Libya's leading publications propagating the *Green Book* internationally, is not in the least discerning in its praise of professional sport, from football to tennis. On the contrary, it makes use of arguments in support of such praise that talk of the 'spirit' of the Olympic Games, and such a position obviously rejects any kind of attack on sport. Thus, elitist sport and public sport would appear to be mutually dependent, and it is revealing that the example of sport in the (old) German Democratic Republic (GDR) – structured as it was from above by laws and decrees, their effects directed downwards in the interests of health, efficiency and excellence in performance – was at one time set up as an ideal fit for emulation (*Med-Post*, 1983).

NOT ONE, BUT AT LEAST THREE KINDS OF SPORT

Libyan publications dealing with sports policy clearly show a similar contradiction to the one existing at the socio-political level (as mentioned above). If one were to enquire about the configuration of sport and physical culture against the background of such contradictions, however, one might find possibilities for resolving the confusions. For in the Libyan people's daily culture, one does not find only one kind of sport, not even a dualism between spectator sport and public sport – one between excellence in sport performance and sport for the masses – that could fit very well the Olympic pyramidal view. Rather, one finds at least three versions of sport running up against one another: the Olympic, public and indigenous games and sports.

Assuming that sports premises show coagulated forms of body culture cast in concrete, one would find that the Western kind of sport executed according to Olympic standards is dominant in Libya as well. The most obvious examples of such premises are the two 'sports cities' in Tripoli and Benghazi, where huge concrete buildings were constructed by Romanian firms according to international norms. Blueprints dated 1969 show construction plans for thirty stadiums, halls, playing fields, swimming pools, shooting galleries and annexes. These plans date from the time when Libya was still a kingdom, and not to times since the Libyan Revolution, as is sometimes claimed. The initial plans, which date from 1964 (followed by additional building contracts since 1965), were enlarged in 1966 when the PanArab Olympic Committee decided to hold its PanArab Games of 1970 in Libya. In 1969, when the Revolutionary Officers put an end to the rule of King Idris, the construction of buildings was already well under way (*Sportstättenbau und Bäderanlagen*, 1969), and so their origin can thus be traced back to the neo-colonial politics of the pre-revolutionary era.

The construction of stadiums still goes on, though, from the topmost Olympic ideal model down to small stadiums all over Tripoli, some sixty-five of which are recorded in a report from the Capital. (The second kind of sport, 'sport for the masses', is of course taken into consideration here.) The department stores, which the Libyan Revolution wishes to see replace the bazaar dealers, also contribute to Western ideas of sport being introduced into Libya. In children's departments, one finds, next to numerous plastic toys, punch balls and garden football sets on sale. This kind of offer neither suits the Libyan environment nor reflects the pronouncements found in the *Green Book*, but it certainly is in the turnover interests of the international market. Pages in Libyan daily papers devoted to sport correspond with those of a Western type of newspaper, as too do the publications of the commercial sports press such as *Afro-Sport*. A National Olympic Committee has been working in Libya since 1962 (*Olympic Review*, 1983), and in order to train highly qualified football players a German trainer from the GDR was hired.

But in Libya one also finds another idea of physical culture different to the Western or Olympic approach to sport based on principles of 'quicker, higher and stronger' as linked to an obsession for measurement, improvement and production, and as assessed in terms of performance data detailing the last centimetre, gram and second. This alternative is bound up with the 'sport of the masses' or public sport, which is organised on the basis of street committees, and has its highest administrative representation in the people's committees for public sport. It receives coverage and publicity through a magazine called *Sport of the Masses (ar-riyadiyya al-gamahiriyya)*, whose overall aims are health, relaxation, education and moral values. Characteristically, one finds cartoons in this magazine, always juxtaposing the handsome, wiry sportsperson with the fat, ugly and inactive individual. Consequently, this kind of sport fits in well with the pioneers' strategies for mobilising the people, which are propagated at the same time as being turned to the end of promoting direct democracy through people's conferences and widespread socialism. It would therefore probably be more appropriate to speak about 'sport for the masses' rather than 'sport of the masses'.

How does this kind of sport become viable? Some of the publicised exercises have to do with Western sports such as football, track and field events, and also gymnastics on apparatus. Their general level of performance does tend to be poor. Beyond the purely numerical grading of such exercises, though, one does observe at sports festivals that both the individual winner and everyone else who has taken part in them are all rewarded with prizes. Gymnastics play a special role here. If health and hygiene are especially emphasised in this connection, it is because these have to do with campaigning for hygienic living. By means of posters, exhibitions and so on, the public, but especially the young, are urged to lead a healthy life – by brushing one's teeth, for example.

In the forefront of public sports activities one unmistakably finds features of political pageantry, and in this respect flags and pictures of Gaddafi play an obvious role (Aljamhiriya, 1979). Musical bands, hymns and fiery speeches accompany the movements. Public gymnastics take place in colourful displays, symbolising changes in arrangements and formations that have occurred on the playing fields. Such a group performing bodily exercises – which are today widespread in countries outside Europe, especially in the People's Republic of China, North Korea, Cuba and elsewhere – can trace its origins to European examples: respectively, to developments in physical culture effected by workers' movements between the two World Wars and to games of the masses idolising Soviet proletarianism (Riordan, 1977). In addition, it should be acknowledged that Nazi Germany took up a similar kind of performance, albeit only temporarily, as seen for instance in the mass games accompanying the Olympic Games in 1936 (Eichberg, 1977). The European socialist sports of the masses were also founded on a co-operation, but partly on competition too, between hygienicism and political pageantry.

ARAB KNIGHTHOOD TRADITION AND BEDOUIN GAMES

In this juxtaposing of Western sport or excellence in sports performance with public sport, however, one still remains far from the special configuration of Libyan reality. This recalls to mind the contradiction between the *Green Book* rejecting elitist sport as a theatrical stadium event on the one hand, and public sport ushering in new forms of pageantry and making stadiums indispensable on the other. But a still closer examination allows a third kind of sport to become apparent: namely, the native Libyan–Arab or bedouin games and physical training. Contrary to the understanding of Olympic officials, the Arabs have not in the least waited all of their history to be 'conquered' by European sport. Indeed, on many occasions in the history of body culture, the Arabs have stimulated Europeans and influenced them, and this has especially been the case in dancing and horsemanship (Hunke, 1978). The morris dance arguably introduced Moorish styles into Europe from the fourteenth to the sixteenth centuries; similarly, many courtly dances, such as the galliard, pavane and sarabande, can be traced to Arabic names, dance patterns and configurations. The 'high school' of riding, which dominated European aristocratic horsemanship from the sixteenth to the eighteenth centuries, is traceable to Arab examples. The horse rider's merry-go-round is of Arabic origin (carousel, *kurradj*), and so is jousting (tournament or tourney, *djarid*). Reminiscences of this kind in Libya induce people to take up some past traditions anew, particularly with regard to further development in riding skills as equestrian sports or horse displays, and also within the military framework of the people's cavalry as well. Gaddafi devotes his extraordinary attention to the traditional riders:

> The preservation of ideals of Arab knighthood and Arab morals is demanded from riders, since they embody all the great ideals of the Arab order of knights, which they ought to pass on to their posterity. The People's Cavalry Troops are a new armed force of the masses. Apart from being a decorative and beautiful picture, they are the foundation of an armed people, whose riders could be turned into armed soldiers and fighters.

> (Gaddafi, 1982a)

But beyond these knightly and internationally influential traditions in Libya, one also finds above all the practices of the broad masses, the traditional games of the bedouin. Whereas colonial ethnography has until recently only given such games a superficial treatment (Mercier, 1927; Gini, 1939), they have of late been receiving a good deal of attention as a crucial part of Libyans' own cultural identity. The games are described and propagated by the magazine *Sport of the Masses*, and are sponsored by an association for promoting traditional Libyan games. As a result, there are more than 150 teams playing the Libyan national ball-game, Al Kora, which resembles hockey. Again and again, one hears repeated remarks about the advantages of playing indigenous games, given the fact that they are especially simple, universally engaged in by the masses with no need for expensive equipment or special dress, and being conducive to girls and women joining in them. A book has been produced describing the bedouin games, and giving practical suggestions to teachers of such games (Mansouri, 1984). Five exercises are dealt with in its first volume: the hockey game, Al Kora; a handball game played in a circle, having no competing sides or winners; a run-and-catch game; broad-jumping over bedouin attire spread out on the ground; and a 'board game' played in the sand. Finally, an imaginary picture is provided of the sports teacher, thus meeting the requirements of the kind of national Libyan leader desired to teach 'sport for the masses'. The book is the result of a research project at the University of Tripoli in which students interviewed older bedouins about their games, and then asked them over to the University for a week to perform their games, discussing with them at the same time whether modifications of the games might be possible.

Has this programme more to it than being merely an aspect of national policy on folklore? Such a possibility cannot be discounted. Recent studies on old games of the masses show how closely their social patterns are related to those of the forms of society in which the games originated (Møller, 1984; Larsen and Gormsen, 1985). Under altered socio-cultural circumstances, old games can, in fact, be looked on as relics or romantic survivals, used for attracting tourists or kept alive to serve political ends. But especially in non-European countries that have been recently experiencing neo-colonialism and the impact of Western sports, other possibilities for development are open. Beyond indigenous, regional, national games and body culture, other forms of resistance and alternatives to new structures of colonialism can surface. In Libya, with its strong bedouin character, indigenous patterns of social behaviour have neither disappeared and nor are they in any way dysfunctional, contrary to what theorists of modernisation often assert. Such patterns could be the basis of a new kind of development that Libya could call its own.

NEO-COLONIALISM AND BEDOUIN SOCIALISM

The three different categories of sport vying with one another in Libya are related to each other in different ways, and they thereby serve as indicators of societal contradictions. Whereas Olympic–elitist sports entail the producing of results and public sports characterise showmanship, possibly of a military nature, indigenous sports reflect aspects of socialisation from below in the shape of enculturation through co-operation (despite elements of competitiveness occurring in individual games). For instance, the winning team in an Al Kora match was obliged to give a party in honour of the losers, which strikingly contrasts with the principle of reward through achievement in industrial societies. Thus, Western sport relates rather to aspects of production, hygienic public sport to reproduction, while the Libyan–bedouin kind of sport – with its emphasis on enculturation – places itself outside production-orientated society. The Olympic kind of sport and the one indulged in for mobilising the masses have this in common: both kinds are superimposed on the people from above, whether by colonial elitism – which was responsible for introducing Western sport into Libya – or by the vanguard of the Libyan Revolution. Indigenous games, on the other hand, are based on the centuries-old experience of Libyan Arabs themselves.

Nonetheless, the main contradiction as apprehended by the people's committees lies between the Western spectators' kind of sport or the elitist kind (on the one hand) and the two forms of public sport and bedouin games (on the other). Gaddafi's interest in this connection appears to be devoted to aspects of bedouin culture, whereas the revolutionary committees give socialist public sport shows precedence over bedouin games. As a result, a hypothetical door is opened enabling one to approach the societal contradictions that were mentioned at the beginning of this chapter. If sport reflects societal structures and processes in their entirety, then reference must be made to the strong position and dynamism of how neo-colonial elitist sport influences the dominant currents of a day-to-day, Westernised neo-colonialisation (itself wrapped up in the automobile society, Western-type high-rise flats, international consumer goods in supermarkets, problems of corruption and pride in economic growth) as permeating self-portrayals of the Libyan state.

Antithetically, the social movement that is being crystallised out at present in the people's committees has to do with the 'mobilisation of the masses', through the framework of the people's congresses, and always referring to Gaddafi and the *Green Book*. At this point, signs of a conflict with groups of the elite becomes apparent, the latter, together with leading circles of the army, being accused of venality. In the forefront of the revolutionary programme one finds an emphasis on both democracy from below and socialism, whereas Part Three of the *Green Book*, which deals with national revolutionary topics, conspicuously receives little attention. This is already obvious in the sartorial appearance of many speakers from the revolutionary committees, who wear the European suit in contrast to the bedouin cloak worn by Gaddafi himself and by the majority of Libyans. One cannot fail to see parallels here with Europe's socialist-inspired 'mobilisation' of the 'political gymnastics show' kind of public sports event and presentation.

BODY, POLITICS AND MYSTICISM

The question arises as to what the third category, that of bedouin games, is signalling socio-politically. Indeed, since 1975, Gaddafi himself has increasingly been legitimating and even celebrating the coherence of bedouin culture by both dressing like a Libyan native and living in a bedouin tent. He meditates in the desert, and has repeatedly stressed its significance for the revolutionary process:

> From the desert, there shines a new era on humanity, the age of the masses. The desert is neither barren nor arid. . . . The desert does not grow grass, but it grows values and immortal cultural messages. Thus history confirms that the desert is a fertile soil, where ideals can grow. But, which desert are we talking about? The Great Desert of the East, the cradle of ancient civilisations and the descending place of heavenly inspirations.
>
> (Gaddafi, 1982b)

At first sight this looks like expressing an interest in folklore, as was the case with bedouin games, but a careful look will show that many aspects of the Libyan Revolution are inconceivable without the specifically bedouin background: indeed, the radical efforts made to rule through democratic councils depend on the bedouin background of self-government at communal level, and the bedouin clash with the bazaar. Their historical background is the social tension between the city – a place of markets and foreign power – and the bedouin countryside. The actual role of the *Green Book* cannot be understood without the traditions in litany and calligraphy used when imparting the message of the Holy Koran to the Believers.

Just as in sport, where Western-inspired spectator sports contrasts with the other two versions (public sport and Libyan games related to each other as a socialist alternative), so a corresponding pattern can be detected in the country's political configuration (see Figure 5.1). The revolutionary committee socialism based on democratic councils and the bedouin way of life are in no way identical: in fact they are rather contradictory, and yet these two different kinds of ideologies definitely appear as connected when contrasted with neo-colonialisation. It is this configuration that gives Libya its peculiarity, and casts light on the internal tensions found in bedouin socialism today. That is why monolithic patterns of interpretation, such as 'military dictatorship' and 'socialist victory', as much as any dualistic interpretation of 'modernisation' versus 'traditionalism', are bound to fail.

Instead, the quotation mentioned at the beginning of this essay – revealing how Gaddafi sees himself – becomes clear, stating that it is the Revolution

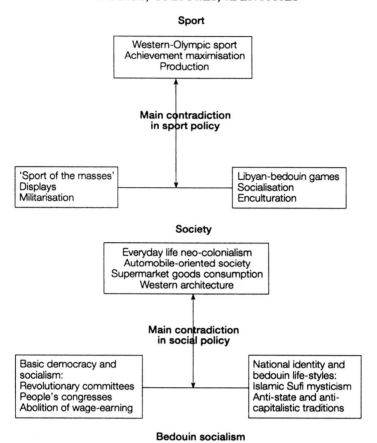

Figure 5.1 Libya: sport and society in comparison

that matters to him, not politics. This claim defies understanding without the historical background to Islamic mysticism, which always attempts to bypass learned authority and directly attain (divine) Truth through sustained concentration and exercise. A special variety of Sufi mysticism was represented in Libyan history by the Sanusi Order, tolerant and not orthodox–inconoclastic, meditative and not ecstatic, political and not quietistic. In Cyrenaica, monastic-like communities had been founded since 1843 and became integrated into bedouin tribes. It was from here that the Sanusi Brethren became politically important, beginning with their struggle in 1911 against Italian colonialism and forming the institutional basis for the founding of an independent Libyan state after 1945 (Evans-Pritchard, 1949; Ziadeh, 1968). The ascetic, and at the same time non-orthodox fundamentalist, features of Libyan radical change after 1969 – with their, somewhat mystical, cardinal concept of 'revolution' – must therefore be seen in this connection. Accordingly, politics seems to be something simply negotiated and abstract, an experts' activity, having little to do with people's everyday lives, whereas 'revolution' is understood as a personal and immediate experience of change and enlightenment – accessible to all – emanating from the desert. There might also be a connection here between body culture and mysticism, but this would be require a special analysis. Islamic mysticism – whether this means the ecstasy of whirling dervishes or the meditation of the Sanusi Sufis in Libya – should, in this perspective, be analysed not only as a religious or political phenomenon, but also as a phenomenon of movement culture.

INTERNATIONAL TENSIONS, OLYMPISM AND A 'THIRD WAY'?

Relationships between sport and society in Libya, and hence between the human body and politics, arguably tell us about more than just the country itself and its three million inhabitants. There are also indications at the level of international sports policies that North–South tensions are growing, and in this connection a growing divide can be discerned between Western Olympic sport, represented organisationally through both the International Olympic Committee (IOC) and other international sports associations, and the representatives of African, Asian, Latin and Indo-American sports (Hietanen, 1982). The latter group of countries command a majority in the UNESCO Commission for Sport, and are increasingly organising themselves around the issue of sport policies within the framework of the non-aligned nations. In such activities Libyan representatives have become especially conspicuous as a mouthpiece for Third World interests (Mansouri, 1981). In this connection as well, sport cannot be seen in isolation. One obviously finds reflected in such activities a new ingredient in the anti-colonial movement that started in Bandung in 1955, but which at that time marked only the very beginnings – and only the 'political' beginnings – of the movement.

After the experiences with neo-colonialisation – with media, consumer goods, multinationals and so forth – what is of great importance at present is no longer the question of political independence, but that of national and cultural identity. In these circumstances, the Libyan experiments in sport could hold out a new fascination for the Third World, since the Libyan experience here perhaps points to a 'third way' between the kinds of sport engaged in by both the East and West bloc countries. This possibility becomes all the more tangible, since two other experiments have in the meantime lost their power of persuasion: both the negative Olympism and the Chinese model. A goal of 'negative Olympism' would be for places like Libya to have their own organisations, run independently of IOC sports, and yet to continue engaging in the usual sports. The Games of the New Emerging Forces (Ganefo), which took place under Sukarno from 1963 to

1966 in Indonesia and during the pre-Cultural Revolution era in China, can be seen as an example of dissociation from Western Olympic sport organisations in the 'spirit of Bandung'. In terms of what was offered in sports content, however, they remained a true copy of the Olympic Games, from the various disciplines contested to the observance of ceremonial, and in consequence no room was left for developing non-Western cultural identities (Ganefo, 1963).

During the Chinese Cultural Revolution, various experiments improved on sports content and began to foster an independent Chinese model: 'Friendship is more important than victory or defeat' (Laage, 1977). It was soon discovered, however, that neither the monolithic claims of 'the new sport' nor the dualistic interpretation of the 'fight of the two lines', nor even the universalising ambitions of Mao's Cultural Revolution, fitted in with the realities of people's body culture. The excesses perpetrated by the Cultural Revolution against China's own culture and traditions, but especially against minorities in regions such as Tibet, undermined the revolution's claim to find a road to people's socialism. Here also, the desire to develop a non-Western cultural identity – or, rather, identities – failed.

Seen in the light of the foregoing assessment, one can find in Libya's sporting developments and contradictions a genuine expression of Libyan experience, thus enabling the country to express its own identity more than do (some) other Third World nations. This could then be a 'third way' that other nations, non-European ones especially, might find attractive. The bedouin socialism of Libya's Revolution is admittedly the distinctive result of its own local presuppositions, and is thus a specific configuration that cannot be transplanted intact anywhere else. It could, nevertheless, act in broad outline as an inspiration for other peoples and places seeking a 'third way', like, for example, Algeria (Sport Sud, 1985; *El Kora*, no date). Even so, the outcome of Libya's internal tensions, as they are embodied by its sporting life, remain open.

NOTE

Translated from German by Jack Sislian, University of Hamburg.

BIBLIOGRAPHY

Ahmad, N. *Die Ländlichen Lebensformen und Agrarentwicklung in Tripolitanien* (Heidelberg, 1969)

Aljamhiriya (The Socialist People's Libyan Arab Jamahiriya) *Revolution d'Al-Fateh en Dix Ans* (Tripoli, 1979)

Eichberg, H. (ed.) *Massenspiele. NS-Thingspiel, Arbeiterweihespiel und Olympisches Zeremoniell* (Stuttgart, 1977)

Eichberg, H. *Sozialverhalten und Regionalentwicklungsplanung. Modernisierung in der indonesischen Relationsgesellschaft (West Sumatra)* (Berlin, 1981)

Evans-Pritchard, E. *The Sanusi of Cyrenaica* (Oxford, 1949)

Fathali, A. and Palmer, A. *Political Development and Social Change in Libya* (Toronto, 1980)

First, R. *The Elusive Revolution* (London, 1974)

Gaddafi, M. *Das Grüne Buch. Die Dritte Universaltheorie* (Tripoli, Bonn, 1980)

Gaddafi, M. 'Speech on the Sixth International Riding Tournament at the Riding School in Tripoli', in *Jamahiriya* (Bonn, 1982a)

Gaddafi, M. 'Speech given at the People's Conference in Sabha, 1977', in *The Road to People's Authority*, in Jamahiriya (ed.) Arab Libyan Popular Socialist Foreign Information (Tripoli, 1982b)

Gaddafi, M. 'Speech before the World Symposium on the "Green Book"', held in Benghazi, 1983

Ganefo 'Documents of the First Ganefo Conference', held in Djakarta, 1963

Gini, C. 'Rural ritual games in Libya (Berber baseball and shinny)', *Rural Sociology*, 1939

Hard, B. 'Interview with Muammar al-Gaddafi', *Information* (Copenhagen), 1983

Hietanen, A. 'Towards a new international sports order', *Current Research on Peace and Violence*, 1982

Hunke, S. *Kamele auf dem Kaisermantel. Deutsch–Arabische Begegnungen seit Karl dem Grossen* (Frankfurt/Main, 1978)

El Kora: Les sports traditionnels en Algérie (no place of publication, no date)

Laage, R. von der *Sport in China, Entwicklungsskizze und Bestandsaufnahme* (Berlin, 1977)

Larsen, N. and Gormsen, L. *Body Culture. A Monography of the Body Culture among the Sukuma in Tanzania* (Vejle, 1985)

Mansouri, A. 'Brief for the Recommendations and Resolutions of the First General Conference for the Ministers and High Officials of Youth and Sport of the Non-Aligned Countries', held in Tripoli, 1981

Mansouri, A. 'The concept of sport in the Third Universal Theory. Public sport', in *International Colloquium on Muammar Gaddafi's Thought in 'The Green Book'* (Tripoli, 1982)

Mansouri, A. *Sport for All* (in Arabic) (Tripoli, 1980)

Mansouri, A. *Sport and Popular Games* (in Arabic) (Tripoli, 1984)

Mattes, H. *Die Volksrevolution in der Sozialistischen Libysch-Arabischen Volksgamahiriyya* (Heidelberg, 1982)

Med-Post, 1983

Mercier, L. *La chasse et les sports chez les Arabes* (Paris, 1927)

Møller, J. 'Sports and old village games in Denmark', *Canadian Journal of the History of Sport*, 1984

Olympic Review, 1983

Riordan, J. *Sport in Soviet Society* (Cambridge, 1977)

Sport Sud *Operation Scorpion* (Algeria, 1985)

Sportstättenbau und Bäderanlagen, 1969

Wright, J. *Libya* (London, 1969)

Ziadeh, N. *Sanusiyah. A Study of a Revivalist Movement in Islam* (Leiden, 1968)

6

OLYMPIC SPORT
Neo-colonialism and alternatives

A CRITICAL VIEW ON OLYMPISM

Since the 1970s the institutional frame of international Olympism has been called into question. The International Olympic Committee (IOC), as an oligarchic, self-co-opting organisation with worldwide monopolistic tendencies, lacks democratic structure, legitimation and control from below. Although a *social* problem from the very beginning, this was not regarded as a special *political* problem as long as the IOC members were, elected or not, a mirror of the nations and cultures represented in the Olympic Games. Since the decolonisation of Africa and Asia, however, and since the rise of non-European sports movements, this has changed, and the Olympic structure now demonstrates a remarkable national-cultural inequality. This has resulted in increasing tensions between UNESCO, where the non-European countries form a solid majority, and the IOC, where Western and European members still dominate. The conflict is even more distinct between the IOC and the non-aligned nations, seventy-two of whom have no representation at all in the IOC. This problem is not restricted to the IOC. It is also a structural problem of the international sports federations in which Western and European functionaries still dominate. It is on this political level that the call for a 'new international sports order' has originated.

But the institutional and organisational question of representation remains a superficial phenomenon. Beneath the superstructure of organised inequality can be found many economic interests wrapped up in sports. Often, one person may occupy simultaneously a top position in an international sports federation and one in the sports equipment industry. This connection allows, for example, a manipulation of the international norms and standards of sports contests, and hence of facilities and equipment, in the interest of the producing firms. Thus institutional inequality results in economic dominance and neo-colonisation by Western capitalism in sport as in other areas.

The economic interests and the facility networks in elite sports are also but a superstructure in relation to sport as a daily activity of the people.

Unfortunately the daily sports of the different peoples in the world have rarely been the subject of empirical studies until recently. The newer research in this field shows an expansion of Western sports into the cultures of Africa, Asia, Indo- and Latin America, at the cost of the national and native games and exercises. Olympism is an important factor in this expansion. It is not by accident that the Olympic Games started at the height of the age of colonialism. The sports disciplines represented in Olympism are – with the sole exception of judo (from Japan) – exclusively of European and North American origin. Missing are, for example, the exercises of the Wushu complex (t'ai chi chuan, shao lin, 'kung fu' and the like), karate, aikido and taekwon do – all from East Asia, but practised today by millions all over the world. Also missing are Thai boxing, the Indonesian martial art of pencak silat, the Southeast Asian ball-games, sepak raga and sepak takraw, and many others. Even Western sports that have become typical Asian mass activities like table tennis and badminton have had problems with Olympic recognition. This lack of recognition of non-Western sports corresponds exactly to the non-recognition of 'Third' and 'Fourth World' nations as members (and a majority) of international sports organisations. The issue at stake is whether sports disciplines should grow out of indigenous national tradition, or continue to be superimposed by colonial and neo-colonial pressures.

The understanding of this problem is also important for avoiding future mistakes when preparing alternatives to a 'Western' Olympic Games. The Ganefo (Games of the New Emerging Forces), for example, arranged in Jakarta in 1963 and in Phnom Penh in 1966, were to be an alternative to Olympism enacted in the 'spirit of Bandung', of anti-colonial solidarity in the age of Nasser, Nehru and Sukarno. But the actual programme of the Ganefo scarcely differed from that of the Olympic Games. So the Ganefo, although an important pioneer project of Afro-Asian independence in sports, failed to develop an alternative to Olympism as a monopolistic force in world sports. An institution opposing neo-colonial structures cannot be effective if it is not rooted in alternative concepts of life, the body and everyday practice.

The single sports disciplines, the individual games and exercises of peoples all over the world, are more than just interchangeable techniques. They are by no means incidental to the peoples and places where they originate and are played. Recent research on behavioural patterns and social configurations in sports shows that important relations exist between the games of a people and their social structures, forms of co-operation and conflict solution, concepts of social space and time, and so on. Thus, Olympic sports are by no means universal but, rather, a specific result of social developments in the frame of European and Western societies over the last 200 years. The configurations of Western sports correspond to the patterns of Western industrial capitalist societies (as well as to those of the East European state economic systems that prevailed until recently).

The following configurational characteristics of Olympic sports, among others, have been discussed in the sociological and historiographical literature:

1. A certain form of competition and fighting, leading to aggression and brutality; often compared to the capitalist principle of competition.
2. The production of results and their subsequent improvement, subordinating the joy of human movement in itself to this goal; often compared to the industrial capitalist (and state monopolist) focus on production growth.
3. The quantification of results, corresponding to the quantification of educational achievement, intelligence quotient, gross national product and the like; this depends on a reduction of the complexity of human achievements and human life, and sanctions the creation of elites and hierarchies, of artificial inequality instead of democratic solidarity.
4. The functional fragmentation and parcelling of space in sport, shaping sport monocultures, standardised sport facilities, separating sport and non-sport activities, men and women, old and young, classes of high and low achievement.
5. The fragmentation of time in sport in accordance with the separation of work and leisure, advancing industrial exploitation and hindering human autonomy and the wholeness of life.
6. Discrimination against women and their subjugation to male patterns of sport, parallel to the male patterns of industrial capitalist (and state monopolist) production.

Such patterns are by no means 'natural' or desirable for all the peoples in the world, so it can be argued, all the more so as they also meet an increasing criticism by the younger generations of the Western metropoles. They have led to consequences that are visible as excesses in world-class sport today:

1. Aggression and brutality as systematic bases of conduct in elite sport.
2. Chemical manipulation of athletes, especially of women athletes.
3. Professional training of children, who are unable to defend their rights and interests.
4. Construction and standardisation of highly specialised top sports facilities that are so expensive that most of the nations of Africa, Asia and Indian-Latin America are excluded from holding Olympics or contests at the Olympic level.
5. The massive appliance of science to top-level sport, so that the individual achievement of the athlete is disappearing behind the achievement of the whole 'system' of trainers, medical personnel, managers, equipment producers and chemists; this competition between complex systems, rather than between individual atheletes, has been called 'the totalisation of sport'.

These excesses are not accidental or marginal phenomena. They are logically related to the configurations of Western Olympic sport as such, and also clearly reflect a part of Western (and East European) societies: 'Quicker, higher, stronger' – centimetre, gram, second. That is why the mere 'purification' of Olympic ideals seems insufficient as well as unrealistic. It is the Olympic principle itself, born at the culmination of colonial history (1896), that is called into question. Olympic sport is thus neither self-evident, nor natural, nor necessary for all peoples. It is socially and culturally relative, tied to a specific historical-cultural formation whose dominance is no longer generally accepted by many, both in and beyond the West. Sport for all, yes – but what sort of sport? This question is open once again.

ALTERNATIVES: TOWARDS A NEW INTERNATIONAL SPORTS ORDER

The age of colonialism has formally come to an end, but the field of sports shows that the mechanisms of international inequality are continuing or even increasing. They function at the political–institutional as well as at the economic level. There are, moreover, initiatives to expand Western sports disciplines into the last village in Africa, Asia and Indo-America at the cost of the native and national physical cultures. But it is not only the age of neo-colonialism today. It is also the age of the masses, of the awakening nations and peoples. The solidarity of Bandung in 1955 was a first signal, but it remained restricted to the top-level political agenda. Now, after the experiences of neo-colonial penetration, new efforts seem to be on the way in the name of 'cultural identity'. Nations and peoples want to take their everyday life, including sport, into their own hands. They are experiencing the conflict between identity and alienation in body culture as well as in daily life. Questions arise about what the alternatives might be to the neo-colonial tendencies of international sports.

In some countries the initiatives towards a new physical culture are arising out of a revival of indigenous and national games and sports. Colonisation, experienced as a subjugation of the bodily self to alien patterns, provokes a new turn towards one's own national cultural traditions. Libyan bedouin games and Chinese Wushu, Indonesian martial arts and rattan ball games, Inuit contests and drum dances in Greenland and Canada: these are just some examples from the three continents. But also in Europe, in places such as Portugal, Denmark and Flanders, the interest in native folk games is reawakening. Even the Council of Europe has taken up the matter. Whether outspoken or not, all of these tendencies develop in the context of social and national criticism of international neo-colonial dominance in sports as tied up with the excesses of industrial capitalism.

Of course, the revival of native games is not without its problems in an era of worldwide capitalist penetration. There is the danger that indigenous games and dances – by being 'developed' – can degenerate into a form of folklore, commercialised and instrumentalised for the attraction of tourists. Games and physical culture cannot be separated from their social conditions (under capitalism), but on the other hand pluralism in the cultures of the world is not possible without there being a variety in body cultures as well. Therefore, the decolonisation of sports can indeed start with the new development of national and native sports and games.

Opposition to Olympic sport also arises in the name of nature. 'Open-air life' is a special Scandinavian tradition, of mountaineering and skiing, while a culture of hiking and cycling in the countryside also developed in the early twentieth century among the German youth and workers' movements. Early English sport was welcomed as 'outdoor games' as long as it was not, in the name of achievement, enclosed by walls for the purpose of standardisation and measurement. Most of the native and national games of the peoples of the world have a close relation to the natural environment.

Such a dimension is reactivated today under the conditions of the threatened environment. It makes a significant difference whether a ball game is played in the open landscape between two communities, promoting adventure in the countryside, experience of the environment and social communication, or whether artificial and expensive facilities with plastic grass are laid out, displaying a stark mono-functionality and requiring environmentally costly automobile traffic provisions and huge parking lots. Thus, the 'green wave' in sport indicates not only an alternative to the dominant Western sports model, with its concentration on results (instead of on adventure and experience) and its consumption and destruction of the living environment, but also indicates a new political and social relation to nature in opposition to industrial annihilation. This means that the 'green wave' in sport and the actual 'green' and ecological movements in Europe, the latter as a political protest against both capitalist and state monopolist growth, have something in common.

A further alternative to Olympic sport is developing in the form of expressive activities such as dance, drama, pantomime and musical-gymnastic performances. Years ago, in the early Soviet Union, the *Proletkult* (proletarian culture) movement tried to develop new forms of expressive physical culture in contrast to bourgeois sport. These experiments were terminated by Stalinism. The new Copenhagen carnival is quite another example of stimulating the creative interest of the masses in expressive body movement, while in other (notably Mediterranean and Latin–American) countries the carnival has a long-standing tradition. African and Afro-American dance forms often merge with this tradition, and they also find a widespread reception in the Western world today. Elements of Scandinavian gymnastics, moreover, are also witnessing a revival in this new context.

However different these forms may be, they have in common their con-trast to the result fixation of Olympic sport and to its 'production' character.

They are focusing on movement and on the body itself instead of on the data produced. Moreover, they further the participation of women, who have had little chance to develop their own physical culture in the frame of male-dominated and male-patterned Western sport. They further creativity in contrast to the fixed forms and standardised rules of Olympic sport. And finally they may – like the old carnival in the European Middle Ages – reveal an element of social protest, a bodily demonstration of the autonomy and the self-liberation of the masses.

Still another form of physical culture arises in exercises of a meditative character. Yoga, t'ai chi chuan and other East Asian exercises, often with elements of Oriental spirituality (Zen, Tantra), are finding a positive reception in the Western world, especially among the younger generations. Native American shamanism too, with its ecstatic and meditative aspects, is meeting new interest elsewhere. These examples show another deficiency of Western Olympic sport: not only deficiencies in nature and expressivity, but also in spirituality. Olympic sport is based on the separation of body from soul or spirit, a separation resulting from very specific Western conditions (which were never undisputed). The meditative exercises deny this separation, and they are also directed against the fixation on results, on quantified sports data and records, which is hindering the inner experience of the body.

THE FUTURE OF OLYMPIC SPORT

All this does not mean that Olympic sport, being in a crisis today, will or should disappear from the scene altogether. Its elements of combat and competition have deep roots in many cultures of the world (although not in all of them). Football or soccer, for instance, has become an important cultural element in many working-class areas as well as in village communities spread across many different parts of the world. It reveals a special complexity and flexibility in adapting to very different cultural conditions, while other Olympic disciplines are not as adaptable. But body culture has always changed throughout history. It will also change in the future. The age of Western colonial dominance is coming to an end, and with it the predominance of Olympic sports. New body cultures will arise (as they have always arisen) from the many and varied cultural traditions of the world.

If democracy is an accepted value among nations, then pluralism and variety in sports will be more appropriate to the reality of a (post)modern world than the uniformity of the principle of Olympic records. Maybe some remnants of Olympic sport will remain as a sort of circus, show-business and media attraction, but they will no longer dictate the exercises in the schools and the games of everyday life. The masses in different cultures, nations and regions will have their own festivals revealing their own patterns, their own traditions, their own historical and future changes.

105

BIBLIOGRAPHY

(While not referenced in the essay, the following items were drawn upon in its preparation.)

Clement, J. 'La force, la souplesse et l'harmonie. Étude comparé de trois sports de combat: lutte, judo, aikido', in Pociello, C. (ed.) *Sports et Société* (Paris, 1981)

Diem, K. *Weltgeschichte des Sports* (Stuttgart, 1971)

Eichberg, H. *Leistung, Spannung, Geschwindigkeit* (Stuttgart, 1978)

Eichberg, H. 'Von der grünen Heide zur fensterlosen Halle. Sport als Landschaftsfrass und Parzellierung des Raumes', *Jahrbuch der Turnkunst*, 1981

Eichberg, H. 'Stopwatch, horizontal bar, gymnasium. The technologizing of sports in the eighteenth and early-nineteenth centuries', *Journal of the Philosophy of Sport*, 1982

Eichberg, H. 'Leistung zwischen Wänden. Die sportive Parzellierung der Körper', in Imhof, A. (ed.) *Leib und Leben in der Geschichte der Neuzeit* (Berlin, 1983a)

Eichberg, H. 'Force against force. Configurations of martial art in European and Indonesian cultures', *International Review of Sport Sociology*, 1983b

Geertz, C. 'Deep play. Notes on the Balinese cockfight', *Daedalus*, 1972

Gormsen, L. 'Sportsudvikling i Tanzania', *Centring*, 1983

Günther, H. *Jazz Dance. Geschichte, Theorie, Praxis* (Wilhelmshaven, 1980)

Guttmann, A. *From Ritual to Record* (New York, 1978)

Heinilä, K. 'The totalization process in international sport', *Sportwissenschaft*, 1982

Hietanen, A. 'Towards a new international sports order?', *Current Research on Peace and Violence*, 1982

Johnson, E. (ed.) *Sport and Physical Education Around the World* (Champaign, Ill., 1980)

Jung, K., Pilz, G. and Jessen, H. J. (eds) *Brutalisierung im Sport* (Westfalen, 1978)

Korsgaard, O. *Kampen om kroppen* (Copenhagen, 1982)

Krippendorf, J. *Die Landschaftsfresser. Tourismus und Erholungslandschaft* (Bern, 1975)

Ku Ahmad, B. and Wong, K. *Silat Melayu: The Malay Art of Attack and Defence* (Kuala Lumpur, 1978)

Lüschen, G. 'Die Nationalen Olympischen Komitees. Konflikte um organisatorische Kontrolle und Verbandspolitik', in Kutsch, T. and Wiswede, G. (eds) *Sport und Gesellschaft* (Königstein, 1981)

Maier, H. *Pa Tuan Chin* (Oldenburg, 1979)

Mansouri, A. 'Brief for the Recommendations and Resolutions of the First General Conference for the Ministers and High Officials of Youth and Sport of the Non-Aligned Countries', held in Tripoli, 1981

Mazrui, A. 'Boxer Muhammad Ali and soldier Idi Amin as international political symbols: the bioeconomics of sport and war', *Comparative Studies in Society and History*, 1977

Mildenberger, M. *Heil aus Asien? Hinduistische und buddhistische Bewegungen im Westen* (Stuttgart, 1974)

Møller, J. 'Sports and the old village games in Denmark', contribution to the Second European Cursus Traditional Sports and Folk Games, Portugal, 1982

Nitschke, A. and Wieland, H. (eds) *Die Faszination und Wirkung aussereuropäischer Tanz- und Sportforum* (Ahrensburg, 1981)

Olivar, C. *History of Physical Education in the Philippines* (Quezon City, 1972)

Pilz, G. *Wandlungen der Gewalt im Sport* (Ahrensburg, 1982)

Renson, R. and Smulders, H. 'Research methods and development of the Flemish folk games file', *International Review of Sport Sociology*, 1981

Riordan, J. *Sport in Soviet Society* (Cambridge, 1977)

Simri, U. (ed.) *Proceedings of the Conference on Physical Education and Sport in Asia* (Netanya, 1972)

Täube, R. *Innere Erfahrung und Gesellschaft* (Frankfurt/Main, 1980)

Tetsch, E. (ed.) *Sport und Kulturwandel* (Stuttgart, 1978)

Sutton-Smith, B. *Die Dialektik des Spiels* (Schorndorf, 1978)

Zurcher, L. and Meadow, A. A. 'On bullfights and baseball. An example of interaction of social institutions', in Lüschen, G. (ed.) *The Cross-Cultural Analysis of Sport and Games* (Champaign, Ill., 1970)

TOWARDS A NEW
PARADIGM

7

BODY CULTURE AS PARADIGM
The Danish sociology of sport

In November 1987 an international meeting at the University of Rennes (Brittany) assembled scholars who study sport and body culture from seven European countries. Among them were ten Danes. It seems to have been the first time that Danish research on the social and cultural aspects of sports had been presented collectively for a public outside Scandinavia. Although the Danish scholars came from different institutions – the University of Odense (Institute of Sports), the University of Copenhagen (Institute of Cultural Sociology), the Danish State Institute of Physical Education in Copenhagen and the Institute of Sport Research in Gerlev – they represented a fairly coherent body of knowledge and of perspective, creating in the foreign audience the impression of a 'Danish School' of sports research.

The Danish contributions were far from unitary, however, and covered a broad range of topics: body and social identity in problematic youth gangs in Aarhus; body culture and sport in the 'Third World' (Libya, Greenland, Indonesia, Japan); the historical study of Danish gymnastics; social patterns in old Danish village games; workers' sports and workers' culture. But whether they were locally or globally oriented, historical or contemporary, they centred around the terms 'body', 'movement', 'culture' and 'society', all seen in a critical perspective. Such a 'Danish Critical School' emerged in the late 1970s, and the materials reported here about this 'school' should be seen as a contribution to the history and sociology of knowledge.

DANISH NON-SYNCHRONICITY AND 'THE POPULAR': THE HISTORY

Danish research into sports rests on a 'real base' constituted by the history and actual structure of sports activity in this country, and this approach has several particularities that make it different from its counterparts in other countries. In comparison with Germany, Sweden, Norway and many other sporting nations, it is striking that Danish sport is not organised in one single sports body, but in four: the Danish Sports Federation (DIF), representing

111

1.5 million members; the Danish Gymnastics and Youth Federation (DDGU), with 0.9 million; the Rifle, Gymnastics and Sport Federation (DDSG&I), with 0.8 million; and the Danish Federation for Company Sports (DFIF), with 170,000. But in contrast to other countries that also have a plurality of sports federations – such as Austria and Finland – the differences in Denmark are not historically rooted in a class conflict between a bourgeoisie and a workers' movement. Rather, the Danish plurality of sports organisations has its origin in another class tension: that between the farmers of the late nineteenth century on the one side and the urban classes on the other. Thus, the DIF represents bourgeois sports but has also included (from 1943) the Danish Workers' Sport Federation, while the DDGU and the DDSG&I both grew out of the farmers' culture.

This situation has not only direct political implications, but also, and especially, it has effects on the level of activity. While bourgeois and proletarian sport from the 1930s onwards differed little in terms of their respective activities, which were primarily focused around competitive sports, the Danish farmers' movement followed from the very beginning a different concept of bodily movement, that of gymnastics. Danish gymnastics had been imported from Sweden in the 1880s. In Sweden – the country of 'the father of gymnastics', Pehr Henrik Ling – as well as in other countries, it never really became popular, whereas in Denmark it did. The Danish farmers and the democratic left-wing majority in Danish politics (*venstre*), standing in sharp political opposition to the dominant minority, the right wing (*højre*), took over Ling gymnastics to demonstrate their own body cultural profile. This happened in contrast to the older military gymnastics of German origin, and also in distinction to the forms of sport that were imported from England to the Danish towns. Compared with these German and English imports, Swedish–Danish gymnastics were to develop a solidarity and community feeling, thereby constituting a special pattern of 'popular sports' (*folkelig idræt*) based in local associations and assembly halls.

It seems that Denmark was the only country where gymnastics really became 'popular' (maybe with the exception of Finnish women's gymnastics). In any case, the term *folkelig* (popular) became a trade mark for this form of physical culture, setting it on equal terms with 'popular' culture in general, which came to consist of popular academies or *folkehøjskoler* (Andresén, 1985), assembly halls in the villages, and rifle associations. All of this was associated with the religious and political ideas of N. F. S. Grundtvig, and also with the farmers' co-operative societies. These movements from the late nineteenth century were based on the religious revival of the beginning of that century, and they united political and spiritual forces, class aspirations and national consciousness, social critique and bodily movement culture. Bound up with the relative delay of Danish industrialisation, there thus resulted a specific cultural conflict situation that shaped a very original development of body culture. Ernst Bloch could have characterised this as a paradigmatic example of non-synchronicity, and something similar showed itself again in the 1960s and 1970s.

From the 1930s, meanwhile, sport had overtaken large parts of Danish physical culture and penetrated both the dwindling farmers' culture and the *folkelig* gymnastic organisations. Also, the most radical of these, the DDGU, had become chiefly a federation of ball sports. This process mirrored the penetrating powers of industrialisation, urbanisation and the capitalist economy in Denmark. Comments could be heard in the 1960s declaring the 'popular' Danish way to be finished, and so the folk academies, the *folkelig* sports and their like were duly condemned to disappear together with their social base, the farmer.

But just at this point in developments, the trend changed. New social movements appeared, giving new impulses to more 'anachronistic' features of the older sports and games configuration: first, there was the youth revolt starting in the late 1960s, largely by students; later, there were the feminist and ecological protest movements, the movement against the affiliation of Denmark with the EU (which did not succeed) and the movement against the establishment of atomic reactors (which did succeed). Parallel to these grassroots movements, the folk academies experienced a boom as never seen before, apparently connected with a change in the social profile of their members from rural to urban. And the term *folkelig* once again became a term of cultural criticism.

In the sphere of sports and body culture, the old contradiction between gymnastics and sports mutated into a new sports critique, and was followed by both new governmental (and social democratic) strategies of sports reform and new practices of alternative body culture. These have resulted in a broad political debate over sport and, last but not least, they have also prompted the genesis of a new type of sports research.

A SPORTS SCIENCE IN QUEST OF ITS PARADIGM: THE FRAMEWORK

Research into sport in Denmark has for many decades been dominated by the approaches of natural science. Unlike, for example, the German situation where pedagogy was the starting point, older Danish sports science tended to focus exclusively on anatomy and physiology. This again reflected, as a sort of superstructure built atop, the base of Danish body culture anchored in the dominance of the old Swedish gymnastics. The Lingian system drew its self-reliance from the conviction of being 'rational' and 'correct' in a natural–scientific sense. In 1909 this had led towards the introduction of gymnastics as a subject at the University of Copenhagen, from which time gymnastics became taught as a laboratory science, as anatomy and physiology. The first teacher was Johannes Lindhard, who became appointed professor in gymnastic theory in 1917. With the physiologist August Krogh,

Lindhard laid the foundations for a research programme of international standing that received worldwide appreciation.

It was in the 1970s, under the changed circumstances of sport and debates about it, that the need for another type of sports research first became articulated in political programmes and then realised in institutional measures. When in 1970–1, a second Institute of Sport was started at the newly established University of Odense, it not only contained natural scientists but also some educationalists. In 1981 the Institute was supplied with a visiting professor in the sociology of sport, and the latest development at Odense moves towards combining the sociology of sport with cultural work on the community level, thus integrating sport (in a practical sense) into 'cultural work in local society'. In about 1982 research on the history and sociology of sport was started at the Danish State Institute of Physical Education. Here a Swedish scholar was invited and historical studies were started, although the possibilities for sports–sociological research remained very limited, the school officially having no right and status to carry out research.

Of greater importance was an initiative that was privately started at the Folk Academy of Gerlev (Zealand) in order to promote the study, discussion and research of social and cultural aspects of sport. This resulted in the foundation of an Institute of Sports Research (*Idrætsforsk*) in 1975, which later gained financial support from the Ministry of Culture. The traditions of the earlier *folkelig* cultural movements (folk academies, Grundtvigian approach, gymnastics) met up here with the new critique of sports and society, resulting in a new frame of reference: a critical study of body culture. This was documented in a series of books and research reports on the history and sociology of sports, but also in texts dealing with the psychology and economics of sports. In 1980, the Institute started a review of sports sociology entitled *Centring*. It also became more and more engaged in research projects commissioned by the Planning Administration, the Ministry of Social Welfare, the Ministry of the Environment, the Ministry of Culture, several community administrations and also the sport federations, especially the DDGU and DDSG&I.

In 1984 a further step was taken at the University of Copenhagen to establish the sociology of sport, and this occurred in the framework of the Institute of Cultural Sociology. A provisional education in the sociology of sports was started, and a plan of studies in 'Sport and Culture' was designed in co-operation with the Institute of PE and the August Krogh Institute. The future of these studies, however, has not yet been decided at the Ministry of Education.

When studies in the sociology of sports started in Denmark in the 1970s, one looked to foreign countries for an orientation. Scholars were invited from Sweden and Norway, articles were translated from American publications, and survey projects were designed along well-known patterns, being hooked up on definitions, questionnaires, statistical descriptions, correlations, discussions of indicators and so on. But this reception quickly led to frustration, since it seemed inappropriate for the expression of a uniquely Danish experience of body culture and sports history. That is why the search for theoretical frameworks turned more towards Germany, where both the neo-Marxist criticism of sport and the systems theoretical approach (Heinemann, 1986) were found to be of interest. Another impact came from sociological studies of the body, its space and time (Elias, Dunning, Eichberg), and this angle was supplemented by an interest in the newer French philosophy of the body and sport, of discipline and postmodernity (Foucault, Bourdieu, Vigarello).

The emerging Danish sociology of sport contrasts strongly with the research profile in other Scandinavian countries, especially in Norway and Sweden. The Swedish sociology of sport is characterised by a strong confidence in 'data' and numbers, in statistics and correlations based on questionnaires. 'We should know more facts' is the classical quantitative and positivist approach, instead of asking 'what are the relations and the cultural patterns of our existing (and not yet existing) knowledge?'. The preferred topic of the Swedish approach is therefore an enquiry into motivations for being active or non-active in sport, as correlated with age, gender, social class and other (mostly rather simple) indicators. At best, the results can show that involvement in sports depends on: the *milieu* where growing up takes place; how one's early experiences in sport lead to positive or negative attitudes towards it; social position; gender; social network; or, in conclusion, the 'total life-situation' (Engström and Andersson, 1983). Results like these may be viewed as trivial: as a social critic in the nineteenth century ironically formulated it, 'poverty comes from poorness'. But they are part of a pedagogical concept of research, the starting point being the question of how to engage as many people as possible in sport, and the result being general propositions for a more intense 'sportisation' of society. Thus the goal is principally affirmative and social–technological, and therefore somewhat removed from a critical analysis of sport and society. Theory is also lacking here to a great extent, the French, German and Danish literature being mostly absent in the references and discussions (Schelin, 1985). The Norwegian sociology of sports is to some extent similar to the Swedish (see Kjörmo, 1979). This might follow from its rather narrow co-operation with the unitary Norwegian Sports Federation (NIF), which corresponds to the Swedish Imperial Sports Federation (RF). Some special emphases have come into Norwegian sociology, however, through a critical sports debate in the 1970s (Bogsrud et al., 1974) and also as a result of the stronger participation of feminists in debate and research (Lippe, 1982). The social critic Johan Galtung (1984) has contributed some philosophical theses that have found international attention, but less so in the sociology of his own country.

The positivistic orientation of the Swedish and Norwegian mainstream sociology of sport has led to a rather wide gap between sociology and history. Between the static and statistical tables of data and the narrative description of historical events or processes, there is very little connection. Swedish sports history is rather well developed, with high-quality academic dissertations and, annually since 1981, the publication of *Idrott, Historia och Samhälle* (*Sport, History and Society*). In spite of this programme, its main stress lies in the collection of historical sources and data, while reflection upon societal relations and theory is mostly lacking. One can notice some dissatisfaction and self-criticism in Swedish sports history regarding this situation, however, and in particular the Swedish interest in popular movements, *folk-rörelser*, could bridge the gap between static sociology and structureless history (Lindroth, 1974; Solberg, 1981). Under the term 'popular movements', sport is treated side by side with religious revivals, abstinence movements, trade unions, political parties, women's or feminist movements and co-operative movements: a combination rather particular to Swedish social history. Another contribution relating sport in society to history has come from Swedish ethnology, notably from the work of Mats Hellspong. And last but not least, there is a young generation of sociology students searching for a more qualitative, more theoretical and more critical approach (Andersson and Gustavsson, 1985).

BODY, CULTURE AND CONTRADICTIONS: THE CONTENTS

How did the Danish school of sports sociology uncouple itself from the American–Scandinavian paradigm? One example illustrates the change in perspective: that of dance. One of the most developed and sensitive studies in the Swedish sociology of sports described dance thus:

> Dance can be pursued as a competitive sport, as motion for health, as social entertainment or as a means of artistic expression. It is especially in this last meaning, as an expression of human creativity, that dance can be said to constitute a special part of body exercises. Its movement is a means of expression, not a method to improve one's health. Dance as a form of art – besides being continuously a bodily exercise in its own right – has also had a great impact on the Swedish women's voluntary gymnastics movement and on sport education in schools, especially for girls.
>
> (Engström and Andersson, 1983)

This description illustrates an approach that has its starting point in societal functions, shifting from these to classify the phenomenon of dance into competition, motion/health, entertainment and artistic forms. This fits very well with the institutional parcelling of industrial society and of bourgeois culture. But does it fit the qualities of dance as the body language of a society or (sub-) culture? What about the magical and cultic or the erotic and psychedelic dimensions of the dancing body? These are the sorts of questions that the Danish sociology of sport is asking, starting with the social aspects of the body. From this perspective, dance is in the first place a rhythm of the body, a structured body-time. Rhythm and record, myth and ritual, word and movement, song and work, feast and drama: these terms hence constitute quite another framework for sociological analysis (and for cultural criticism), as picked up on in one of the foremost Danish books on gymnastics, sport and society (Korsgaard, 1986a).

From this perspective it becomes clear that the functionalist approach makes important dimensions of social sensuality disappear: neither the drum dance of the Inuit nor the disco dance of industrial youth culture will fit into the narrow pattern of this approach. Nor is it just the art of ballet that is foremost in representing the human creativity of dance; and nor are individual 'creativity' or artistic 'expression' in the tradition of ballet the universal characteristics of dance as bodily exercise. Rather, monotonous repetition, social vibrations and joint rhythm can constitute dance as a social structure, embodying collective social identity and creating altered states of consciousness, social ecstasy and joint trance. Focusing on the body and taking this as the starting point thus leads to quite another sociology of sport, one that almost inevitably includes an in-depth critique of stale positivistic categories.

The body is therefore the starting point for Danish sociological studies of sport (Bonde, 1986; Korsgaard, 1982, 1986a). There are of course other approaches, especially those of organisational sociology or of conflict theory (Klausen, 1988). But it is no accident that the journal *Centring* has the subtitle '*krop og sport*' ('body and sport'), or that the Danish Society for the History of Sport bears the subtitle '*krop og kultur*' ('body and culture'). By putting the body into a central epistemological position, the Danish sociology of sport his avoided (or tries to avoid) becoming one of the many 'hyphenated' sociologies that developed from positivistic sociology, especially in the 1950s and 1960s. Hyphenated sociology (as in sports-sociology) means that sport is treated as if there was nothing specific in its field, or, in other words, as if it were without the body. Sports activity or non-activity (when correlated with age, gender, social class and the like) can here apparently be treated in much the same way as any other activity – say, collecting stamps or watching television. The central point of the activity, the body, is thereby systematically excluded from sociological analysis, and in so doing assigned to the natural sciences, anatomy and physiology, just as the involved mind is assigned to psychology and the dimensions of change to history. The interrelating connections of the activity – sport as fascination, sport as culture – disappear. (The same has happened, for instance, with the sociology of art and the sociology of music: degenerating towards a technique that simply measures people's visits to museums or attendance at concerts, while

117

expelling the arts and the music themselves from sociological analysis, allocating them to the purely aesthetic disciplines.) The body as an epistemological starting point shows that this type of sports sociology is a hyphenated science that leads away from the topic itself, whereas the Danish type of sports sociology of the body is trying to break this restriction.

The notion of the body in such a context could be fundamentally misunderstood as being just a biological phenomenon: rather, the body is cultural. With its background in the historical Danish *kampen om kroppen*, the struggle about the body (Korsgaard, 1982), Danish sociology cannot but concern itself with the social plurality of body and sport and their many historical changes. It is multiplicity and change in the context of social processes that constitute the 'culturality' of the body. Body culture (or movement culture) is thus becoming a new term standing side by side – or even above – the terms of sport, *idræt* (an old Nordic term), gymnastics, play/game and dance.

Body culture as the new paradigm places sport in the context of culture. This constitutes another discourse, unlike accounts of sport in terms of system and function, of stratification and correlation. Sport as culture fits well into the allocation of sports in recent Danish policy, sport having been administered under the Ministry of Culture since 1976. (Previously it was under the Ministry of the Interior.) Increasingly, sport is regarded as a part of 'cultural work', especially in community research and community politics, and this development also fits in well with the merging of the sociology of sport and cultural sociology at the University of Copenhagen. Cultural sociology was established in Denmark in the early 1960s, in contrast to the quantitative and non-philosophical sociology of the American type. Unlike at the prevailing positivist-oriented Institute for Sociology, cultural sociology here tried to combine interdisciplinary aspects of sociology with psychology, history and anthropology, all this under the heading of 'social critique' (Due and Madsen, 1983; Eichberg, 1987).

This culturalist approach means that the sociology of sport is becoming part of other cultural studies such as those concerned with youth culture and subcultural studies, gender studies and 'Third World' studies. Youth culture research – having some points of contact with the Birmingham School – has newly developed its own interest in sport, dance and the staging of the body (Bay *et al.*, 1985–7), and gender studies have started in Denmark to incorporate feminist debates about female sport. While the Norwegian discussion has put higher priority on woman's equality in sports (Lippe, 1982), there has been a feminist tendency in Denmark to criticise this approach as comprising a homogenisation and assimilation on male premises. The question of how a more female–feminist body culture could be developed also relates to the history of Danish gymnastics, given its strong early female participation. Yet it has remained an open question. It gives inspiration to a line of corresponding research directed at males, however, and to studies tackling the gender politics of the masculine sporting body (Bonde, 1986). 'Third World' studies have placed sport in the framework of a neo-colonisation of everyday life in non-European countries such as Indonesia, Libya, Greenland and Tanzania (Larsen and Gormsen, 1985; Eichberg, 1988), and this research is accompanying attempts to come up with a new exchange of culture at the grassroots level, especially between villages in Sukuma (Tanzania) and the DDGU. In Norway, Galtung (1984) has supported this dialogue with some more general ideas about the crisis of Western sport, and about what is involved in learning from non-European cultures.

Thus the political dimensions of sport are being treated centrally in the Danish sociology of sport, contrasting with purely pedagogical or social–technological perspectives in positivist research. 'Political dimensions' means taking the differences and contradictions of sports as a central theme. The Danish 'Critical School' does not mean to formulate an alternative programme as such, but systematically to shed light on actual contradictions and historical changes. One field of these contradictions is the local community. Sport in local communities, in both urban municipalities and in rural districts, has been treated in very different ways. Questions asked in various municipalities have included:

1. What are the patterns of sport involvement? (Jespersen and Riiskjær, 1980–4).
2. How do sports facilities mirror and determine the physical and social space of sport?
3. What are the political contradictions between elite culture, the old popular culture (including, among other things, sport) and a new popular culture (including grassroots movements)? (Skot-Hansen, 1984).
4. Can new activities be introduced to promote social and cultural aims (Povlsen et al., 1987)?

In 1987 the Ministry of Social Welfare also started a whole series of projects all over the country, with the goal of testing sport as a means of 'social prevention'. Sociologists of sport were asked to take part in the evaluation, and to accompany the practical work by active research. This may result in new inspiration and possibilities over the coming years. The Social Ministry's request is an indicator of the growing interest of different state administrations in sports research. Thus, studies have been ordered by the Ministry of Culture (Nielsen and Riiskjær, 1983), by the Planning Administration (Riiskjær, 1984b, 1986), by the Ministry of the Environment (Eichberg and Jespersen, 1986) and by the Ministry of Education.

In contrast to both the local community approach and the requests from different ministries, the impact of the large sport federations on research in Denmark has not been as powerful as in other countries, their competing influences balancing each other out. But the federations nevertheless do require more and more research. The DDGU has drawn upon sociological expertise more at the grassroots and debate level, for example, favouring the topics of sport as cultural exchange with the 'Third World' (Larsen, 1983) and popular (*folkelig*) sport in relation to Danish identity (Thygesen and Rasmussen, 1984). The DIF has worked more centrally, producing a noteworthy report displaying remarkable vision (DIF, 1983). The DDSG&I has requested a survey from the Institute of Sport Research in order to define their own profile and image for the future (Riiskjær, 1985a), while the DDGU and the DDSG&I together have ordered a report on the problems of recruiting permanent trainers to local clubs (Berggren *et al.*, 1987). Sports federations, local communities and the state have been the main actors in supplying sports with financial means, with the remarkable increase of inputs coming from the state during the last decade. Since the 1970s this has been a major issue in Danish sports, and as a result one object of research has concerned the economic dimensions of body culture (Klausen, 1988; Riiskjær, 1984a, 1984b). The newest changes involve the impact of the private economy on sports: sponsorship, professionalisation, the advertising industry and the mass media (Møller, 1986; Albret and Møller, 1987). Will this capitalisation and mediatisation also bring about a change in body culture, and what type of change?

Last but not least, the contribution of the Danish sociology of sport has focused on the social ecology of sports, and on the relations between the body and social space – or, to be more precise, between sport facilities, nature, planning and culture. This interest has led to studies of the 'green waves' in sport, that is to say of changes in how sports relate to nature in the open air (Eichberg and Jespersen, 1986; Jespersen and Olwig, 1986). This has resulted in a criticism of the traditional planning of facilities and of the dominant 'container architecture' of sports (Eichberg, 1987; Riiskjær, 1985b, 1986). Its latest pay-off has been an architectural competition for an alternative sports hall, taking particular account of feminist considerations (see Chapter 4).

THE LIVING WORD AND THE 'TRIALECTICS' OF SPORT: THE METHODS

Since the 1970s it has not been enough for the Danish sociology of sport to go its own way in terms of content, since its methods have also had to be questioned. Several attempts have been made to use the traditional methods of questionnaire and statistical correlation in order to describe the new experiences (e.g. Berggren *et al.*, 1987), but the results were not satisfying, and the really enriching insights and discussions of these studies came more from outside their formal methodologies. Research has therefore turned increasingly to less conventional sociological methods: participant observation, introspection, in-depth interviews and action research. Here sociological work meets the methods of cultural anthropology and ethnology, with

120

psychology and psychoanalysis also being consulted on occasion. It has been a favourable coincidence that this combination or synthesis already possesses a basis in Danish sociological research, in the form of cultural sociology as an integrating discipline. In the anthropological method of cultural sociology, in action research and in 'dialogical science', the sensual principle of communication is re-actualised: and it is precisely sport and body culture, the sensual dimensions of social life, that call out for this method of exploration.

But the dialogical science developing here has roots going somewhat further back into the Danish history of knowledge. At the beginning of the nineteenth century, the poet and religious 'shaman' N. F. S. Grundtvig formulated a vehement critique of existing educational practices and dominant ways of thinking. His criticism became the starting point of the folk academies, and also the intellectual material for the so-called *folkelig* (i.e. popular) movements in Danish politics, religion and culture. Central to Grundtvig's critique was the rejection of the 'black school', which produces and demands reproducible knowledge. In contrast to 'dead knowledge', Grundtvig developed the idea of 'the living word', which claims that all true education should be based on the sensual medium of reciprocal learning and teaching. All state schools, subjecting children to authority and to the dominant elites, should disappear. They should be replaced by 'popular enlightenment' (*folkelig oplysning*) in the folk academies, in free schools of young adults who are reciprocally communicating with each other. Though this revolution in education never took place as such, the folk academies did come to play an important role in Danish cultural life, the 'living word' becoming or remaining a permanent utopian vision of non-authoritarian communication (Korsgaard, 1986a).

The newer methods of exploration and contents of research, both moving away from positivism, have become interrelated with a method of interpretation that has also been changing. The common patterns of sociology, including the neo-Marxist 'critical school' of the 1960s, have been characterised by the dominance of dualistic or dialectical interpretations and forms of questioning. Sports activity or non-activity, emancipation or suppression by sport, elite sport or mass sport, sport oriented towards competition or sport oriented towards social solidarity, professional sport or amateur/leisure sport? These are the elements featuring in some of the classical discussions, always being dualistic in their configuration. Danish sociology, with its historical background in a greater variety of sports alternatives, has experienced again and again the limits of these approaches and has therefore developed another method of questioning: the so-called 'trialectics' of body culture.

The starting point in the 1970s was the dissolution of the unitary concept of sport as just 'one sport'. The latter notion had been established internationally since the 1930s, and had determined until the 1960s the picture of sport as a pyramid. Sport had been regarded up to this point as an interconnected system of hierarchical levels with different grades of achievement.

'The top makes the breadth, and the breadth bears the top' was a slogan of this pyramidal concept. On all levels the same principles were operating: competition and achievement – 'faster, higher, stronger' – coupled to the measuring of performance in centimetre, gram and second (c.g.s) or in points scored/awarded. The characteristic questions were, for example, why are people active or non-active? How do they reach a high degree of achievement, or why do they remain on a low level? How is sport in different societies well developed or under-developed? How does the modernity of sport reveal itself, and how is traditionalist opposition against sport manifested? All of these questions were characteristically dualistic in form.

Since the end of the 1960s, and beginning with both the neo-Marxist critique of sport and the appearance of social democratic reform strategies, the picture had been changing. The unitary pyramid of sport began to be regarded as less than convincing. Should sport in schools and in leisure contexts really and necessarily follow the same patterns of achievement as do top sports? The answer seemed to be 'no', since there exist two principally different patterns: top sport and mass sport, sport of elite competition and sport for all. While top sport organises itself in pyramids, all participants are more or less equal in sport for all, equal in relation to their wants for health and recreation. The characteristic questions became: is one particular phenomenon a part of competitive sport or of leisure sport? What is the fundamental difference between capitalist and socialist sport? How does sport on a high performance level relate to the social hygienic values of sport? What is freedom and what is repression in sports? Again, though, the concerns here came to be expressed in dualistic questions. In Denmark, the result was sports legislation breaking down sports into two parts: a set of laws for elite sports in 1984, and a set of laws governing sport for all (which is in preparation). But this partition into two never completely fitted with the specific Danish tradition of *folkelig* sport. The 'popular' body culture had always been far removed from elite sport, without – given its political and body cultural alternatives – just being hygienic sport for all. Was it, then, a third alternative?

The question has been reinforced by the 'new waves' that have spread in Danish movement culture since the 1970s and that have involved yet again neither elite nor mass sport, but have embraced expressive activities (with a gymnastics background), meditative exercises, body therapies, old and new games and 'green' outdoor activities. All of this has resulted in a more systematic sociological exploration of the 'Third World' of sports: the result being the 'trialectic' approach. By this is meant:

1. The pattern of *achievement sport* is constituted on the base of production: producing results, measuring, comparing and raising standards. The dynamics involved in this model are continuously dominant in sport as well as in society at large, working in the direction of maximisation,

expansion and hierarchialisation. The relevant experts are top athletes, trainers and even scientists of 'anthropomaximology' (a Soviet term for the science of top performance), from the physiologist to the psychologist. The social ecology of this type of sport consists of standardised spaces for sport, made uniform all over the world and split up into the specific monocultures of the different branches of sport.

2. The second model is the pattern of *fitness sport*, a social–hygienic 'sport for all' or welfare sport. Here the values are health and well-being in their physiological and social–psychological meaning, striving for 'natural life', motion and pedagogical integration. The experts here are the physicians and the pedagogues, both guaranteeing the rules of welfare (the right ways of moving and nourishing, upright poise and social adjustment). There are also relations existing between the straight line of this social hygienic view and the straight line of hygienic sports space: clean and clear to survey, panoptical; the order of the gymnasium, of the fitness centre and of the school.

3. The third model, in comparison, is not that easy to survey. Here the relevant aspects are *body experience and social sensuality*. They find their expression in the older forms of popular sport as well as in dance and in the 'new waves' of body culture. This is the field of surprise and laughter, of non-steered historical change and its subversive qualities. Experts giving general orders and instructions have no place here, yet one finds specialists of another type: the master and the juggler, the acrobat and the buffoon. The space of this type of body culture is characterised by a culture of curved lines: outside on 'the green' or in 'the street', the labyrinth of dance, the derelict industrial plant, and spontaneous sport in the space just around the corner, in nearest proximity to daily living (see **Figure 7.1**).

THREE SCENARIOS AND THE EMERGENCE OF HYBRID RESEARCH: THE PROBLEM

The 'trialectics' of body culture could be misunderstood as a description of the landscape of organised sport in Denmark, with the DIF standing for the achievement type, the DFIF for the fitness model and 'popular' sport (DDSG&I and DDGU) for the third possibility. But this would mean neglecting the methodological aspect of 'trialectics'. Every real, concrete phenomenon is blended and hybrid, demanding a differentiated analysis. The 'trialectic' method places at our disposal only a set of categories, not a scheme of contents. Top-level achievement sport, for example, also produces (but only secondarily) body experiences that during recent years have been described as 'sport as yoga'. The real phenomena are neither clear in classification nor neatly 'trialectic', but the categories and the methods of interpretation can – if 'trialectic' in form – produce new and critical insights.

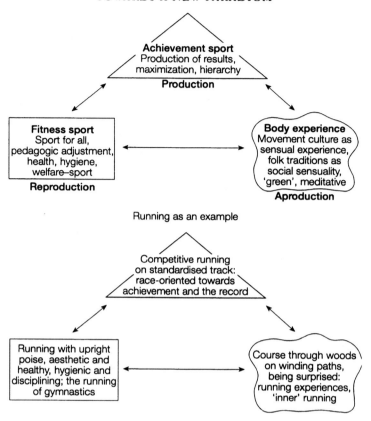

Figure 7.1 A 'trialectic' of sports

This proves to be true not only in relation to sports. Some years after the first 'trialectic' analysis of sport, the Association of Danish Communities published a study that presented ideas on the future of community planning (Kommunen, 1987). In contrast to earlier discussions, which had also often followed dualistic patterns (speculating about capitalist or socialist futures, right-wing or left-wing programmes), the new study proposed three scenarios: the efficient, the consensual and the decentralised society:

1. The *efficient society* will get its dynamics from the strength of international competition. This presses in the direction of producing results and organising society on the basis of individual career planning, of a consciousness of quality, of a high degree of information in the combination of production and consumption, as well as of electronic data technologies. Faith in technological solutions and enthusiasm for achievement, service orientation, modernisation, centralisation and 'de-ideologisation' characterise the associated system of values.

2. The *consensual society*, on the other hand, is not convinced that technology and growth are enough to bridge the gap between the new classes in society. New potentials for conflict arise, demanding an elaborate set of organised balances, a harmonisation of interests and a permanent 'deal' about maintaining consensus between the three social partners: business, labour and the state. This involves constraints prompting thoroughgoing co-operation both between community and the state and between different political wings: all in all, the upshot is to force the partners towards a consensual 'centre'. The plurality of life styles is accepted, but at the same time the individual may experience a feeling of helplessness and weakness *vis-à-vis* the powerful system of consensus, which invades deep into everyday life. Co-operation is here developing in the direction of a corporate society.

3. The *decentralised society* is, again, more eccentric than the centre-seeking corporatism. It stresses quality of life and holism, everyday democracy and coherence of life. This means a high priority for the family and neighbourhood, for the environment and closeness, and here the community is gaining functions that the (central) state is losing. What is lost by de-specialisation can be gained by a grass-roots movement. This does not mean equalisation, and still less the imposition of uniformity by force, but does entail an acceptance of differences in life styles. Simultaneously, however, it also demands a 'flatter' structure of representation (instead of steep hierarchies) and a new cultural community identity.

This triangle of the efficient, consensual and decentralised society scenarios shows clear parallels with the 'trialectics' of sport. This does not necessarily mean a direct determination of one model by another, but a correspondence between the one set of relations (achievement orientation, corporatism and grass-roots orientation) and the other set (achievement sport, social hygienics and social body experience).

'Trialectic' differentiation therefore opens a way to critical analysis, both of sport through society and of society through sport. 'Critical' also means that the sociology of sport, as well as general sociology, can thematicise its own problems with a measure of 'distance' from the materials under study. Swedish sociology has self-critically reflected upon its own situation as 'hybrid research under liberal corporatism' (Elzinga, 1982). Is this true for sports research too? The Swedish research is indeed undergoing a general process of integration into administrative sectors, and it is thereby becoming part of liberal corporatism; it is fitting in with the new alliance of different classes, their democratic political representations and the state bureaucracy. The result is a 'hybrid scholar', being partly researcher and partly administrator, and he or she is already selecting research aims with 'scissors in the brain'. One implication is hence that censorship from outside will not be necessary to gain political control. This rather bleak picture, which is true not

125

only for Sweden, calls for alternatives. The classical alternative would have been free academic research or self-conscious political engagement and affiliation of research. 'Trialectic' reflections as well as empirical evidence have shown that this might be thought of as too dualistic a proposal. Hybrid research, with its background in the corporate system, does not have only one alternative, just as state sport versus 'free' sport (in the meaning of the free market) does not provide one with the whole picture of societal alternatives.

The Danish sociology of sport is part of the Scandinavian system, and it thereby reflects in part the corporate tendencies of Scandinavian societies. But it has simultaneously an historical background in the *folkelig*, those popular movements with their decentralising traditions of the folk academies, grass-roots orientation and cultural criticism. Could it be this tension that has advanced and promoted the development of a Danish 'Critical School' of body culture in the last decade? But this tension is always unstable. It is on the one side liable to the changes of the *zeitgeist*, the fluctuating spirit of the age and of fashion. This is true especially in 'postmodern' times, with their quick consumption of 'paradigms', and with the very concept of paradigm change itself becoming a fashionable 'paradigm'. The Danish tradition of 'the popular', of '*folkeligheden*', can thereby become part of a new consumer society, living from the consumption of critique and 'paradigms', and with the body culture paradigm being just one of these. On the other hand the balance of critical research in Denmark is linked to the tendencies of 'hybrid research', to the all-encompassing Scandinavian model of the '*behandlerstat*', the state of social hygienic welfare and treatment. Is sport on the way to becoming a part of this corporate and authoritarian treatment system? And will sociological research accompany, describe and – precisely by means of its critique – come to promote this process?

BIBLIOGRAPHY

Albret, P. and Møller, J. (eds) 'Showsport', *Centring*, 1985
Andersen, M. (ed.) *Splinter af dansk idræt* (Odense, 1988)
Andersson, J. and Gustavsson, K. *Har idrotten någon mening?* (Örebro, 1987)
Andreséan, A. (ed.) *The Danish Folk High School Today* (Copenhagen, 1985)
Bay, J. *et al.* (eds) *Ungdomskultur. Årbog for ungdomskulturforskning* (Copenhagen, 1985–7)
Berggren, F. *et al. Kommerciel gymnastik* (Odense, 1984)
Berggren, F., Ibsen, B. and Jespersen, E. *Idrættens trækfugle* (Vejle, Kastrup, 1987)
Berthelsen, A. *Frispark* (Copenhagen, 1983a)
Berthelsen, A. *Den røde idræt* (Odense, 1983b)
Bogsrud, N. *et al. Idrett for alle* (Oslo, 1974)
Bonde, H. *En stålsat karakter i et hærdet legeme* (Copenhagen, 1986)
Bouchet, D. and Engelbrecht, J. *80's børn og unge* (Odense, 1984)
Brinch, J. *et al. Fysisk kultur og sport i Kina* (Gerlev, 1980)
Christensen, K. *Fodboldspillet. Teori, historie og fascination* (Aarhus, 1983)
DIF *Idræt 80-udvalgets rapport* (Copenhagen, 1983)

BODY CULTURE AS PARADIGM

Due, J. and Madsen, J. *Slip sociologien løs. En invitation til 80'ernes sociologi* (Copenhagen, 1983)
Eichberg, H. *Die Veränderung des Sports ist gesellschaftlich* (Münster, 1986)
Eichberg, H. *Hvem er vi da i grunden?* (Copenhagen, 1987)
Eichberg, H. *Det løbende samfund* (Gerlev, 1988)
Eichberg, H. and Jespersen, E. *Den grønne bølger* (Gerlev, 1986)
Elzinga, A. 'Forskningspolitiken och den liberale korporativismen', *Sociologisk Forskning*, 1982
Engström, L. *Fysisk aktivitet under ungdomsåren* (Stockholm, 1975)
Engström, L. and Andersson, T. *Idrottsvanor i ett utvecklingsperspektiv* (Stockholm, 1983)
Fasting, K. *et al. Idræt/motion some trivselskabende faktor* (Oslo, 1984)
Galtung, J. 'Sport and international understanding. Sport as a carrier of deep culture and structure', in Ilmarinen, I. (ed.) *Sport and International Understanding* (Berlin, 1984)
Heinemann, K. *Grundbog i idrætssociologi* (Aabybro, 1986)
Hellspong, M. *Boxningssporten i Sverige* (Stockholm, 1982)
Jespersen. E. (ed.) 'Psykologi og sport' *Centring*, 1985
Jespersen, E. (ed.) 'Religion og sport', *Centring*, 1986
Jespersen, E. and Riiskjær, S. *et al. Idræt i lokalsamfundet*, vols 1–4 (Gerlev 1980–4)
Jespersen, E. and Olwig, K. (eds) 'Friluftsliv og natursyn', *Centring*, 1986
Kjörmo, O. *Bibliografisk oversikt over idrettssociologisk litteratur for tidsrommet 1966–1978* (Oslo, 1979)
Klausen, K. *Per ardua ad astra* (Odense, 1988)
Kommunen *Kommunen i 90'erne* (Copenhagen, 1987)
Korsgaard, O. *Kampen om kroppen* (Copenhagen, 1982)
Korsgaard, O. *Kredsgang: Grundtvig som bokser* (Copenhagen, 1986a)
Korsgaard, O. *Krop og kultur* (Odense, 1986b)
Larsen, N. (ed.) *Ulandsarbejdets kulturelle dimension* (Vejle, 1983)
Larsen, N. and Gormsen, L. *Bodyculture. A Monography of the Body Culture among the Sukuma of Tanzania* (Vejle, 1985)
Lindroth, J. *Idrottens väg till folkrörelse* (Uppsala, 1974)
Lippe, G. (ed.) *Kvinner og idrett* (Oslo, 1982)
Møller, J. *TV-sport* (Copenhagen, 1986)
Nielsen, K and Riiskjær, S. *Eliteidrætsudøvernes vilkår* (Copenhagen, 1983)
Povlsen, J. *Den befriede tid – den forbandede tid* (Odense, 1983)
Povlsen, J. *et al. Fritidsarbejde i nærmiljø* (Odense, 1987)
Riiskjær, S. (ed.) 'Sport og økonomi', *Centring*, 1984a
Riiskjær, S. (ed.) *Idrættens omstilling* (Copenhagen, 1984b)
Riiskjær, S. (ed.) 'Arkitektur. Krop og sport', *Centring*, 1985a
Riiskjær, S. *Idræt i opbrud* (Gerlev 1985b)
Riiskjær, S. *Idræt, kulturpolitik og planlægning* (Copenhagen, 1986)
Schelin, B. *Del ojämlika idrotten* (Lund 1985)
Skot-Hansen, D. *Kulturpolitik og folkultur* (Copenhagen 1984)
Solberg, G. *En bibliografi över folkrörelseforskning 1930–1978* (Stockholm, 1981)
Thygesen, J. and Rasmussen, O. (eds) *Idrættens hovedbrud* (Vejle, 1984)

8

A REVOLUTION OF BODY CULTURE?

Traditional games on the way from modernisation to 'postmodernity'

In 1989 dramatic events changed Eastern Europe, but the popular revolution was not restricted to the political level. In sport and youth culture, body culture and everyday life, change had already been on the way. Three events from 1989 provide an indication of the variety of movements for change.

'Qilaatersorneq – drum dance' was the headline of the leader in the Greenland *Atuagagdliutit-Grønlandsposten*. The newspaper was commenting on the meeting of the Inuit Circumpolar Conference (ICC) in Sisimiut (Greenland) in the summer of 1989. For the first time in history, such a conference united Inuit (Eskimo) from Greenland, Canada, Alaska (USA) and Soviet Siberia. This could be seen as a step forward in the 'Arctic Revolution', a turn towards decolonisation. The newspaper's leader underlined the dramatic, sensational and historical character of this political event, but it also referred explicitly to a body cultural event: the drum dance. The Siberian Inuit came carrying their drums. This was significant because the dance had been banned for several decades as being reactionary, 'religious' and 'separatist'. Now drums sounded through the whole week of the conference, moving the participants to tears. The meeting culminated with music and dance as its focal point, uniting for the first time people from all parts of the Inuit world in their common traditional ceremony.

The drum dance is the ancient centre-piece of Inuit festival culture, a dance of laughter and enjoyment, play and display, as well as an instrument of conflict resolution (by duel dancing) and a technique of shamanic ecstatic healing (Eichberg, 1989). In the summer of 1989 the drum dance appeared in its new political context, being a corporeal manifestation of modern national identity, of Inuit identity. In the festival it appeared also in its new musical context, alongside keyboard songs, Country and Western music, choral singing in the Christian missionary tradition and, last but not least, rock music. Yet it was the drum dance that moved the conference participants to tears and marked the emotional mood and significance of the Inuit meeting. Through drums and dance, collective identity was expressed corporeally.

128

DRUM DANCE, *SOKOL* AND A DEFEAT OF SPORTIVE NATION-BUILDING

Another event from 1989 is revealing. In May, Polish political activists met to consider the re-establishment of the gymnastic organisation known as '*Sokol*'. The Polish *Sokol* (Falcon) movement was founded in Galicia in 1867 as a part of nationalist and Pan-Slavic gymnastic movements in several Eastern European Slavic countries. Later it became a base for the National Democratic Party in Poland, and in the inter-war period it was a sort of para-military corps as part of the extreme right. Thus, classical phenomena of both Polish nationalism and Polish body culture re-emerged from the historical underground.

After 1945 the *Sokol* had been prevented from reappearing. Its property and its facilities were transferred to the state sports federation or to communal cultural centres. But things changed in the years of martial law following the *coup d'état* of 1981. In 1984 the authorities in Krakow registered a '*Sokol* centre'. The extent to which this resulted from the nationalist opposition press (from below) or from the military patriotic stabilisation strategy (from above: from the side of the Jaruzelski regime) is difficult to know. Maybe the different interests mixed and united in the common project to counteract the crisis of the so-called 'August generation' that was characterised by political resignation, deterioration of health conditions in the context of ecological constraints, psychological depression and, in general, 'social pathology' (Bogusz, 1988). Five years later, it was possible officially to re-establish the gymnastic federation. Again, nationalism and body culture, historical reminiscence and political change were all linked together.

And there is a third case. When in October–November 1989 the East European revolt reached the German Democratic Republic (GDR), the universities reacted quickly by declaring important changes. The chancellor of the Humboldt University in East Berlin expressed the need 'to butcher the holy cows whose death had been demanded for several years'. He summarised this change by referring to three measures: the obligatory Russian language course was abolished; the courses in Marxist–Leninist ideology were supplemented with other philosophical contents and were no longer awarded grades; and sport was dropped as an obligatory subject and became voluntary (Hass, 1989).

The role of sport as one of the 'holy cows' – side by side with the language of occupation and the state ideology – is not surprising and can be understood against the background of the traditional sport politics of the Communist Party. Through the visibility of its international sports results, the 'socialist nation' GDR was to prove its existence, and the object was hence nation-building by the means of sport. What failed in other fields – democratic consent, economic production, cultural glamour – succeeded at the level of sport, and was quantitatively visible in the Olympic ranking list.

129

Thus, sport became a sanctuary of state-patriotic effort, protected by a strict taboo against serious criticism. Only in fiction such as the novels of Erich Loest (1978) and Christoph Hein (1982) was the critique of sport articulated. But there had grown a popular aversion to the politics of sport, breaking through finally in the revolution of 1989 with sharp attacks against the privileges of the sporting elite. The butchering of the 'holy cows' at the universities made clear that the critique was directed not only against top sports. As with the Polish *Sokol* and the Inuit drum dance, the ties between the national question and body culture became visible, but in distinctive forms. Both the attempt at nation-building from above as well as its defeat under the pressure of a popular movement manifested themselves in sport.

NATIONALISATION AND BODY CULTURE: A HISTORICAL PANORAMA

The relationship between national identity and body culture is not a new phenomenon, but nor is it ancient, archaic and universal. It can be documented in a series of structural changes and innovations occurring since the emergence of modern popular nationalisation in the late eighteenth century. A number of cases can be identified.

1. The German *Turnen* gymnastics appeared in the years after 1810. In the early years and up to the revolution of 1848 it constituted the broadest mass base of the German nationalisation, outnumbering by far the choral societies and the organisations of student nationalism (Schröder, 1967).
2. Among the Slavic peoples of Eastern Europe, the German *Turner* movement gave rise to corresponding national gymnastic efforts. The *Sokol* movements started among the Czechs in 1862 and took over several forms of exercises, but in so doing turned gymnastic nationalism against its German political content (Blecking, 1987).
3. Zionism and Jewish gymnastics were also linked to each other from their start in the 1890s. Under Nazi repression, it was especially Jewish sports that maintained a protective network for the German Jewish community (Bernett, 1978).
4. Ling gymnastics in Sweden is often regarded as a nationally neutral and strictly scientific, physiological and anatomical model of exercise. In fact, it arose around 1800–10 in the context of the Nordic Gothic movement, the Swedish version of intellectual national romanticism (Sandblad, 1985).
5. In Denmark the gymnastic movement was related to the romantic nationalism of quite another social class. Inspired by the Swedish Ling gymnastics, it started in 1884 on the basis of both the farmers' democratic nationalist network and their Grundtvigian folk academies (Korsgaard, 1982).

6. In Iceland it was a national youth movement that, from around 1900, essentially contributed to the development of sports. An important role in this process was played by the ancient *glima* wrestling style (Einarsson, 1988).

7. The different forms of Celtic wrestling are not only related to Icelandic wrestling, but their renaissance and transformation was, in a comparable fashion, linked to regionalist and nationalist cultural movements. This link is particularly documented by the 'Druid', Dr. Cotonnec, founder in 1928 of the Breton *gouren* organisation (Jaouen, 1985). In Brittany indigenous wrestling has pushed traditional games into the background of nationalist attention, but their significance is rising (Floc'h and Peru, 1987). The same is true for *soule*, the traditional Breton ball-game (Moëlo and Le Bihan, 1986).

8. In Ireland, on the other hand, while Celtic wrestling disappeared, it was Gaelic athletics and the game of hurling that attracted nationalist emotions. From 1884, the Gaelic Athletic Association became the base of mass militant Republican resistance to British colonisation (Hörbe, 1983).

9. The Scottish Highland Games possessed, from 1819 onwards, a certain significance for the rise of the new Scottish nationalism. At the same time, though, the Caledonian Games in North America showed that this type of sportive festival culture did not automatically lead to an ethnic–political nationalisation (Redmond, 1982).

10. In the Valley of Aosta, traditional cast and bat games were organised in sports clubs that, from 1920, contributed to the ethnic identification of this Franco-Provençal minority in Italy. *Tsan, rebatta, fiolet* and *palet* thus constituted a body cultural equivalent to the valley's political autonomy, realised after 1945 (Godio, 1985).

While nationalist and regionalist sports movements have, in most of these cases, tended towards a sort of uncoupling, emancipation or separatism, forming their own sportive identities on a smaller scale, in English sport the reverse is found. Indeed, the relationship of English sports – mainly ball-games (notably cricket and rugby) – with British colonial expansion, imperialism and upper-class education has been fully documented (Mangan, 1986).

Over a period of 200 years we arguably see a general picture of rising nationalism tied into growing national sporting differentiation. Hence, industrial modernity is not only characterised by a universal standardisation and homogenisation of sport and body culture, corresponding to the homogenising effects of the industrial system (Eichberg, 1973, 1978, 1986), but at the same time it does support a counteracting, subversive tendency towards multiplicity and heterogeneity – one that breaks through especially in social–historical situations of change and unrest, such as surfaced in the early nineteenth century and then again in the years between 1900 and the 1920s.

DIFFERENTIATION OF FOLK SPORTS: AN INTERNATIONAL PANORAMA

Is all of this just a historical reminiscence? The revolution in Eastern Europe, with its breakdown of the colonising state sports system, provides another answer. Today the political situation needs to attend to the dialectics of uniformity and multiplicity in a new way. Also, from this perspective, some tendencies in Western Europe in the 1980s can be seen in a new light.

Starting at the international level, one can observe a new wave of international competitions and sports festivals starting in the 1980s. In May 1985 the First Games of the Small Countries in Europe was arranged in San Marino. Under the sponsorship of the International Olympic Committee, they attracted participants from San Marino (20,000 inhabitants), Monaco (25,000), Liechtenstein (26,000), Andorra (38,000), Iceland (240,000), Malta (330,000), Luxemburg (365,000) and Cyprus (632,000). Among the 'giants' of these games, therefore, were nations that have little chance of excelling at the Olympic Games. But in other respects – in the sport disciplines and patterns of competition, and also in symbols (Olympic rings), ceremony and ideology – the small states' games mimicked the Olympics. One month later, in July 1985, the First Inter-Island Games took place on the Isle of Man. The participants came from islands such as Åland (23,000 inhabitants), Anglesey (68,000), Faeroe (44,000), Gotland (53,000), Guernsey (54,000), Iceland (240,000), Isle of Man (65,000), Isle of Wight (120,000), Jersey (76,000), Malta (330,000), Orkney (19,000), Shetland (8,000) and St Helena (5,500). Again, there was international standard sport on the programme (such as athletics, cycling, soccer, swimming and volley-ball), along with Olympic-type ceremonies and medal tables. However, the traditional Manx dances at the opening ceremony and the games' symbol – the ancient Celtic *triskell* – indicated another direction of possible development.

Also in 1985 the First Eurolympics of Minority Peoples was hosted by Friesland in the Netherlands. Some 400 participants came from Brittany, Cornwall, Elsass-Lothringen, Belgian and French Flanders, Friesland, Lapland, the Molucs and Räto-Romania, all these being peoples without a state of their own. Here too gold, silver and bronze medals were awarded, but the practical activities consisted of three different types. These were: (a) classical standard sports (athletics, soccer, volleyball); (b) traditional sports and games from the participants' ethnic cultures (Celtic wrestling, Frisian regatta); and (c) cultural competitions (singing). These games thereby included an alternative view on the multifarious character of sports culture. This approach was planned to continue in 1988 with the First Spring of South European Games, planned for Corsica. Participants from France, Greece, Italy, Portugal and Spain were invited to show off their indigenous traditional games and sports, including a tambourine ball game from Italy and Languedoc, *tsan* and other bat games from the Valley of Aosta, water jousting from Lyon and

Givors, stick fights from Portugal and pelota from Euzkadi (Spanish and French Basque regions). But in spite of the sponsorship of UNESCO and the International Fund for the Development of Physical Education and Sport, this alternative type of sports and games festival was blocked at the last moment by changed political circumstances.

At this stage of sports festival 'development', points of contact turn up with both festivals of folklore and international festivals of dance and music. Among these, the Inter-Celtic Festival in Lorient has had special significance for sports. This started in 1970 as a week of dance and music in Lorient (Brittany), with participation from all of the Celtic nations and in recent years also from Galicia (in Spain). Later, Celtic wrestling and traditional Breton competitions came to play an increasingly significant role. It was perhaps this type of cultural festival and arrangement that UNESCO hoped for with its Major Programme of Education, launched in 1983, demanding initiatives on three levels:

- sport: national (state) strategies and plans for raising the standards of attainment;
- sport for all: promotion of health, fitness, recreation and training for democracy, in co-operation with the World Health Organization;
- traditional games and dances: protection and development of national heritage and cultural activities, implemented by the initiative of youth movements.

Although considerable emphasis was placed on the reappearance of traditional games in these festivals, this should not conceal that Olympic-type sport – producing results under standardised conditions – is still the dominating pattern.

Paralleling these international competitions and festivals, a new form of international research emerged in the 1980s. One of the first initiatives came from a French–Danish–German network of researchers, known as the Institut International d'Anthropologie Corporelle (IIAC), which arranged its first seminar in 1987 in Brittany; and this initiative was followed by others in Denmark, Germany and France (Eichberg, 1989). In November 1988, a seminar under the aegis of the Council of Europe in Vila Real (Portugal) looked at 'traditional games' and produced plans for a general inventory for their registration (Dufaitre, 1989). On a regional level, a conference about Nordic Sports assembled researchers from Denmark, Finland, Faeroe, Iceland, Norway, Saamiland and Sweden in Denmark in 1988. Here, as under the IIAC, more theoretical and political questions about the relations between body culture and national or regional identity were a focus of interest. The same interest may have influenced the development of sports geography, appearing as a new sub-discipline in the 1980s (Bale, 1986).

SOME NATIONAL CASE STUDIES

The parallel international competitions and the international research interests are neither accidental developments nor independent variables. They form a kind of superstructure above a range of societal dynamics and problems that could be observed at national and regional levels. In Eastern Europe, the Soviet system was based on a general disrespect for popular games and folk sports. These were regarded as being:

- in conflict with social and economic modernisation generally;
- linked with religious cults or celebrations;
- bound to ethnic nationalism or separatism;
- pre-industrial and archaic–ritualistic, bound to their origin and pattern of life; and
- far from the result-related functions of modern sport, such as fitness, discipline and work co-operation.

There seem to have been only two exceptions where traditional games and sports were tolerated or promoted, and these were:

1. When they could be accepted as a preparation for international top-level sport. In this way, traditional Trans-Caucasian wrestling was used as preparation for Soviet *sambo* and international judo. Several world champions, European champions and top athletes on the Olympic level developed in this way (Chatschkowanjan, 1972).
2. When traditional games from Russian ethnic culture were accepted and transformed into organised sport disciplines, included in the state-wide classification list, and streamlined with unified rules, federations and championships of their own. *Gorodki*, a throwing game in the bowls or curling family, was sportised in 1923, and *lapta*, a ball-game related to baseball, was similarly 'upgraded' in 1953 (Riordan, 1986/7).

The recent transformations in Soviet society have taken place in parallel with alternative movements in the world of sport, especially in the field of youth culture. One can expect that some special Russian traits will emerge, as well as some exotic imports, especially from East Asia (Riordan, 1989). In the non-Russian republics inside the former Soviet Union, the consciousness about one's own national body culture has never disappeared, and will revive along with the wider emancipation process. Examples include the traditional games and sports of Georgia, *chidaoba* wrestling, riding games, archery and others (Natschkebiya, 1964). Similar tendencies have been reported from Kazakhstan. Traditional games of riding, wrestling and archery, related to nomadism at the crossroads between Mongolian and Turkish cultures, have to some degree disappeared as a result of social changes during the century. Others have been repressed because of their relations to shamanism or to 'separatist nationalism'. Recent developments, however, have favoured a renewed practice of the old games, for example in connection with the New Year celebration (*Nauryz*) in March. But a new interest is also being shown in 'erotic games' such as *ak suiek* (white bone), in which two teams of young people, girls and boys, try in the warm summer night to find a bone that is thrown by a referee as far as possible into the steppe. The one who finds it runs back to the referee, shouting loudly, but the rival team members try to take the bone away from him. Both teams get involved in the fight for the possession of the bone. Some participants, however, are searching for other experiences and get lost in pairs in the vast steppes. Evidently, such traditional games are related to emotions that have survived their changing societal 'functions'.

In Western Europe some parallel processes have occurred, and the examples reported here appear to demonstrate some very different body cultural configurations existing as initial patterns of national distinction. The Basque games and contests form a special pattern of strength competition, for instance, and have developed reciprocally with the political strife after national autonomy (Franco, 1978). In the Valley of Aosta, as already noted, bat games are fundamental as part of an identity-related sport culture (Daudry, 1990). In Flanders distinctive games from the towns' inn culture, as well as children's games, have developed towards a national profile. Academic research, having a long tradition in this field in Flanders, has rather concealed the connections between these recent transformations in body culture and the political dimensions of national identity (Renson and Smulders, 1981). In the case of Brittany, there are detailed descriptions of how traditional games and sports, together with traditions of music, dance and festivity, constitute an essential factor of economic development. The language and the body together are here fundamental for regional and national economy (Denez, 1988). Portugal, being one of the very few European nation-states without national or ethnic minorities, has at the same time been the only state in Europe to develop initiatives for research into – and the practical promotion of – traditional games (Soares, 1980). National conservative as well as popular communist incentives are working in this field, converging and struggling against each other.

In the Nordic countries, traditional games have been examined both by an older positivist research (Stejskal, 1954) and by recent ethnology (Hellspong, 1989). There has been very little effort to relate these activities to modern sport and modern cultural identity, with the exception of Icelandic *glima* wrestling. Recent research in Denmark – combined with the practice of a 'living sports museum' – points in a new direction, however, and this is one where the historical traditions of body culture are transformed into actual practical alternatives for the present (Eichberg, 1985; Møller, 1990). Among the Saami (Lapps), by contrast, it is Western sport that has formed the base for national separation in the running of their own organisations rather than in their own types of activity (Pedersen and Rafoss, 1989). The Saami traditions are instead significant for other fields of culture, in jazz and rock music, in theatre and in healing, where Saami shamanism reappears as body cultural practice. Similar developments have happened with Greenland's Inuit culture. The Greenland Sport Federation (GIF) is an independent organisation, but remains a copy of its big colonial brother, the Danish Sport Federation. Knowledge about traditional Inuit games is used by the GIF only as historical reminiscence and legitimisation, without any practical relevance. It is more in the field of dance, music and theatre that the traditions here have found a new field, mostly in the drum dances and mask games. The famous Tuukkaq theatre, founded in 1975 in Jutland and later transformed into a 'Fourth World' theatre, the Silamiut theatre, started in 1984 in Greenland as the national theatre, and the Aasivik summer festivals of music, dance and culture have played an important role since 1976 in this transformation (Lynge, 1981).

FOURTH-WORLD AND ETHNIC MIGRATION: NEW TRIBES

In Inuit North America similar tendencies have emerged with different cultural effects. On the one hand, the Northern Games (or World Eskimo Indian Olympics) and other forms of competitions have both revived indigenous practices and sportised them to a certain degree. But these games have not only included 'disciplines' such as tug-of-war, the one-foot high kick and canoe races, but also more complex cultural contests such as drum dancing, igloo building and tea boiling (Ipellie, 1979). On the other hand, old native American and Inuit games have been introduced into schools, and some Alaskan elementary schools use them for physical education purposes. With games such as leg wrestling, one-foot high kick, two-foot high kick, stick pull and toe kick, teachers try to reach several pedagogical aims: the result is more creativity in physical education, more fun, skill and strategy components in school sports, and a more efficient development of multicultural awareness and understanding. The paradox lies in the search for 'new games' by discovering something old, or in a breaking away from what have become 'traditions' (those of modern sport) by turning towards tradition (the indigenous traditional games). The paradox is of a fundamentally pedagogical character. The games are treated as isolated, culturally neutral and transferable elements of the curriculum, while 'space does not permit a description of their respective heritage or customs' (Frey and Allen, 1989). As implied, a related field of revival in North American Inuit body culture is music and dance, especially drum dance. These are revitalised in the context of a growing ethnic awareness (Johnston, 1978), but – and in contrast with the drum dance of the Siberian Intuit, which gained its special dynamic under circumstances of interdiction and suppression – one can observe some traits of museumisation and folklorisation in North American drum dance. The structural traits of sportisation, pedagogisation and folklorisation are in no way restricted to Inuit societies. They demand some theoretical in-depth study, and we shall return to this later.

It would be an error to presume that the issue of traditional games as bound into national or cultural identity in sports is simply restricted to 'marginal' or 'traditional' peoples. (But note that the Inuit are 'central' to their own world, whereas Paris and Frankfurt are 'marginal', and that they are living in a twentieth century no less archaic than that of the West Indians in London.) To further illustrate the modern (if not postmodern) dimension, attention must be paid to emerging ethnic identities resulting from migration. Legacies of colonial domination, capitalist labour market strategies and other mobility factors have created new ethnic situations in many Western European metropoles. This new ethnicity finds its visible expression not least on the level of body culture. But here positivist research is mainly concentrated on methods of measurement, of quantifying levels of 'integration', instead of structurally analysing the processes of identity and alienation. The sociology of sport has, for example, researched the integrative effects of 'ethnic sport' among Turkish migrants in West Germany, producing highly elaborate statistical details and calculations but *without* mentioning which body cultural traditions, sports and games the Turkish migrants carry in their cultural heritage (Frogner, 1985). Islamic body culture, with prayers, hygiene and taboos, Turkish oil wrestling, dervish trance dancing and other practices rich in their gender aspects, spiritual dimensions, historical transformations and ethnic cultural distinctions, are systematically neglected by the positivist sociology of sports. Studies of the subcultural dynamics of Asian and African immigrant dance have been more illuminating. Indian Punjabi *bhangra* dance in London suggests that the new social, technological and commercial situation of the migrants produces (or is creatively counteracted by) new forms, based on ethnic traditions, but resulting in a new type of body culture (Hargreaves, 1989). In addition, West Indian crowds have changed the character of English cricket, making it a less subdued and more flamboyant milieu.

Another aspect of national or ethnic distinction in modern or hyper-modern sport, as just implied, lies in the activities of fans. Football and ice hockey have for some decades been the scene of escalating fan conflicts, especially in Eastern Europe – in some cases approximating to anti-Soviet riots. In March 1969, for example, the fan enjoyment surrounding the Czechoslovak team's victory over the Soviet Union in the ice hockey world championship exploded towards rebellion in several Czech and Slovak towns. Soviet barracks, Aeroflot offices and other symbols of Soviet occupation were stormed or set on fire. It is the English hooligan who has become the internationally most notorious phenomenon of fan violence, however, and his chauvinist appearance has been documented in detail (Dunning, 1988; Williams 1985). But the national aspects are also visible in his contrasting counterpart, the peaceful Danish *roligan* (Peitersen and Kristensen, 1988).

Both migrant subcultures and fan subcultures show that the connections between ethnic distinction or national identity (on the one hand) and body culture or sport and games (on the other) are not at all restricted to residuals from 'traditional' societies condemned to 'marginal' situations. Attention to 1980s postmodern or hypermodern tribal tendencies also concerns the Western metropoles of international standard sport. In this connection, even modern and standardised American baseball has become analysed and visible as a 'national sport', whether regarded in terms of family structure, gender balance, violence control and 'civilisation' (Zurcher and Meadow, 1967) or as an expression of modern American 'rationality', quantification and spatial organisation (Guttmann, 1978, 1988). But could this game or its social configuration become, sooner or later, simply historical: as historical as the WASP (White Anglo-Saxon Protestant) hegemony in the United States?

PAST AND FUTURE, STYLE AND IDENTITY

Starting from the Soviet Union in the East, a superpower under the impact of national revolutions that were accompanied by revolutionary changes in the fields of sport and body culture (Riordan, 1990), one can finish in the West with the other superpower. Its 'ethnic revival' will not be restricted to linguistic aspects, to the expansion of Latino Spanish, to black English and other pidgins in the United States. All in all, the panorama shows that the revival and modernisation of traditional games and sports is part of a much more extensive societal process, one related to historically established and actually transformed connections between body culture and ethnic (or national) identity. Does this signify a body cultural revolution on a large scale? And, if so, what has tradition got to do with this revolution?

Some assumptions about the evolutionary linearity of history have to be revised. Andrzej Wohl, the Polish Marxist scholar, having analysed sports history all of his life from the paradigm of evolutionary progress, has recently reflected on the phenomenon of new games in the 1970s–1980s as a paradox of the 'old' and the 'new':

> To some extent it resembles games and play in small medieval towns when the townsfolk were not yet differentiated in terms of wealth and prestige. Such holiday games and entertainment were noisy and merry and the entire urban population took part in them. But this only appears to be a return to the past; it is, rather, a look at the future world, once again integrated, though on a new and different basis than in the past.
>
> (Wohl, 1989)

The dialectics of past and future is evidently one conclusion to be drawn from the transformations in the 1980s, but this possibility must be examined in more depth.

Another conclusion is that the survival of traditional games and sports as 'revival' depends on much more than simply their style. Such differences in 'style' from one region or nation to the next, as is often stressed in publications about Celtic wrestling, are interesting as a visible mark of distinction. But it is not the different style as such that makes Celtic wrestling an 'ethnic' or a 'traditional' sport. What is important is the social and cultural context, rendering a certain practice of sports or games a bodily expression of identity. The focus on context and identity, on social practice rather than on just style, on the complexity of a body cultural configuration, makes the study of traditional games more difficult. But is it indispensable for evaluating the historical change of traditional games. And it is the precondition for developing them towards an alternative to sportised standardisation, towards the practice of survival. It is at this stage of reflection that historical study (within the perspective of cultural sociology) can assist us in better understanding the demands of practice. Let us look at one example in more detail.

RAISING THE CROSS

In 1777 the rector of the parish of Commana (Brittany) wrote a report on some strange, but usual, forms of contest and competition in his town. Every year, a religious procession was held when heavy crosses and church banners were carried along a traditional route:

> As soon as the crosses and banners are outside the church, and when they have begun their way, the young people flock together around them and everybody wants to carry them. They are quarrelling, they are pulling them from each other, they are even fighting in view of the Holy Sacrament which is carried in the procession. And in order to get more freedom for their activities, the young people withdraw as far as they can from the clergy, who sometimes lose sight of them. Then they dispute, they laugh high and loud, they try the banners and crosses, they let them down and bear them in this way or let them fall, causing cries of scorn, by this. The clergy are forced to leave their places to prevent the disorder. These are the processions of Commana and of Saint-Sauveur looking like – which being uninformed – one would think to be popular riots.
>
> (quoted in Peru, 1986)

And, so the rector went on:

> It is the weight of the crosses and banners which makes the glory of the carrier, letting them down in the described way. That is why the perches of the crosses and banners can never be large and heavy

enough for the fancy of those who want to carry them. And one has even some times discovered that they have put stones between the textile of the banners in order to make them heavier.

(quoted in Peru, 1986)

The report continued by complaining that even children, who are sometimes too weak, take part so that the crosses fall down and are damaged. After returning from the procession other competitions occur inside the church, where the carriers try to carry the crosses and banners as low over the ground as possible.

The respectable people deplore this disorder; the political authorities have condemned and banned them by their decision from May 26 and August 17 last year. They have ordered and prescribed the church-warden in charge to appeal even to the authority of parliament in order to stop this.

(quoted in Peru, 1986)

But there arose a conflict between the clergy and the elders of the parish. The clergy decided to lock the banners away while the elders insisted upon their traditional ceremony with banners and crosses as usual. This resulted in a confrontation, with

the carriers of the crosses and banners leaving the church in spite of the clergy's interdiction and, without doubt, by order of the elders or at least by one of them who drove the people out of the church, urging them with a loud voice to abandon the priests and to follow the signs with which some inhabitants started the tour around the churchyard. The priests, frightened by this riot, fled into the vestry. One can well imagine the triumph of the carriers of the crosses and banners when displaying them and letting them down on their way out of the church.

(quoted in Peru, 1986)

CONFIGURATION: THE BODY'S RHYTHM, SPACE AND ENERGY

The complaint from Commana described vividly the complexity of popular competition that developed towards what is today regarded as the traditional game of raising the perch (*le lever de la perche*). This configuration of body culture, visible in 1777, can be basically described in the following terms.

The *time* of the pardon competition in Commana is the time of festivity, of celebration (the rhythm of repetition). Compared to modern sport, what prevails is not the streamlined and 'futurist' time of record production; no 'faster, higher, stronger', and no expectation of future improvements on the record. There is no training for tomorrow either. The game is decided here and now – and next year as it has been last year and from time immemorial.

140

The *space* of the pardon game is the environment of the townspeople, of the living and the dead. There is no need for a separate, specialised facility for competition, not even for barriers to keep the public at a distance, and so the church, the churchyard and the road passing the town and the landscape form an ensemble of spatial identity. The game is taking place where people are 'at home'. Moreover, the game is in itself part of a social process establishing 'home': that is, securing a spatial identity bound up with the extent and surroundings of the village.

In the centre of this time–space configuration it is the *body* and its abilities, its strength and skill, that forms the object of public interest. The competition of the strong bodies, the 'strong men', is more important than the spiritual contents as they are defined by the clerical authorities. The religious celebration is at the same time a spectacle of the body, displaying its capacities. But the body raising the cross cannot be understood in its narrow physical meaning: the body here is more than what is beneath its skin, and in this case it is not least a body of laughter, of vibration that is collective, as the report tells with indignation. The body is not only strong, it is also grotesque. The convulsive body, laughing and laughable, is at the centre of this popular carnivalism. The pardon game is also working with vibration in another dimension, with music. The competition would be fundamentally misunderstood if we did not hear both the ringing of the bells, fastened to the heavy banners, and the songs of the procession. Competing and singing are hence combined. Their combination – walking rhythmically, bearing heavily, singing loudly – makes up the collective bodily vibration, and it is this that comprises the 'material' base of religious ritual and cult called spirituality. There is no singular 'belief' commanding the social process but, rather, the collective body laughing, singing and in common vibration. Perhaps the most fitting word for this occurrence is the body's *energy* (Eichberg, 1987).

The body cultural process constitutes a certain *inter-personal relation*. In the case of Commana this is principally a competitive relation: competition between individuals and, as we know from other cases, competition between collectives, between town quarters, streets or villages. The competition also has a gender dimension: it is an exercise for the 'strong men'. Women are not mentioned, and maybe this reflects the state of Catholic patriarchy in contrast to what has been described as the 'Breton matriarchy' (Audibert, 1984). Historical sources reveal that Breton women also took part in wrestling: on the other hand, there have also existed 'men's games' under matrilineal and non-patriarchal conditions. Last but not least, we should not forget that women are present both in the procession and in the dance, following or related to the celebration. The procession is just one part of a larger body cultural arrangement and this festivity is in itself not only a competition between the carriers of the banners; it is also a sort of dance, a common rhythm, a music of body movement. Maybe the complaining rector could have said something about the erotic aspects in the rising of the banners, but his Catholic decency prevented him from doing so. Moreover, the inter-personal relation is characterised by the fact that as well as 'strong men' participating, there were also children trying to lift the heavy burden. The participants are not, as in modern sports, sorted according to classes of weight or strength, and instead their game relation is a form of togetherness.

Is any *objective* registration of results found in the competitions of raising the banner? Compared with measurement in modern sport, they are not objectified, since there is no record or quantification. The banners are made heavy by stones, but their weight is neither measured nor systematically compared or raised towards a record. On the other hand, there is perhaps some other sort of objectivity, a verdict here and now as to who can raise the cross or carry the banner all of the way. But this decision is related to human beings, not to abstract results. It relates to concrete, present strong men and (eventually) to their memory. No quantitative ciphers will come out of it, but the event can run into a story, an oral tradition about a successful (or unsuccessful) somebody: *la gloire de les porters*.

The bodily process – time, space, body, energy, relation, objectivity – is related to a superstructure of values and measures, all describing and pre-scribing the ideal pattern of the game. This *world of values* can extend from the simple scale of decision or measurement up to the complex mythology of the game and its relation to the meaning of life: it is where Christian and pagan values are conflicting, as was so clearly the case in Commana in 1777.

Another superstructure arching above the body cultural process is the world of *institutions*, regulating and controlling the 'correct' execution of the game. Where modern sport has produced written rules, referees, control-ling bureaucracies, and the like, the raising of the banner in Commana is based on local self-determination and *ad hoc* control of the public, and maybe also on the informal level of self-organisation and age-grouping of the youngsters. In this world of institutions, political conflict arises, confronting the participants and their elders with (on the one hand) the clergy, including the reporting rector, and the political authorities (on the other hand). From this dimension of the configuration, so the 'popular riot' is erupting.

THE MODERNISATION OF GAMES

It is seductive to relate the 'riot' of Commana to the political and social revolution that erupted eleven years later in Brittany and some months later still at the French national level. There can be little doubt that the little story of 1777 illuminates one of the sources of the great Revolution of 1789. But seen from the complexity of body cultural configurations, the Revolution of 1789 was not just a realisation of the traditional game against the clergy. Rather, and on the contrary, the body cultural revolution taking place from the late eighteenth century onwards did away with the traditional games, replacing them with new configurations: by sport, gymnastics and folklore.

This revolution of body culture fundamentally changed the configuration of games and exercise. The raising of the banner, for example, would (just like other traditional games) be marginalised or disappear, or it could be 'preserved' as folklore, or it could be sportised. A combination of folklorisation and sportisation can hence be seen in the 'raising of the perch' in Brittany today (Peru, 1986). Can the modernisation of the old games thus now be described within the categories of configuration (as introduced above)?

Modern games have uncoupled their activities from the rhythm of celebration. The *time* pattern of festive repetition has been replaced by a new type of time order: leisure time as the opposite of working time. For competition sport, a new type of continuity has been introduced and directed into the future: training in order to raise the record, growth in achievement, increase and maximisation (*citius, altius, fortius*). For this purpose, the perch must be measured and supplied with a graduated weight.

The *space* for the game now becomes more and more specialised and standardised, and in the process isolated from the environment of everyday life. The church and churchyard, which the authorities since early modern times had tried to 'cleanse' of pagan games and pleasures (the case of 1777 was just one of these), was now effectively restricted to the 'sacred' ceremony. The road was taken over by traffic, later by automobilism, and it largely disappeared as the space of games. The Breton *gouren* is also an example of how the management of space was changing: in premodern times there had been a 'fool' or 'police', holding the public at distance by grotesque measures, by a whip or a frying-pan, black and blackening by soot, but now this human shield has been replaced by material barriers. The stadium, sports hall and highly specialised facilities are now the final result of this process.

Concerning the *body* itself, interest in strength has diminished and has been replaced by the fascination with speed and velocity. The old games of strength and endurance become marginalised, or they are transformed into quick and dynamic exercises (like football and boxing). It is the streamlined body that is now required. In the new body cultural configuration there is no longer a place for laughter and popular carnivalism: the fool disappears and the music that had accompanied running contests, wrestling and other games is replaced by silence. The body's *energy* gets transformed into what is now regarded as a modern dynamic.

The focal point for the new dynamics of sport has mutated into the production of results, and this constitutes the new form of *objectivity* in movement culture. Movement and activity have been subordinated to achievement: quantified, measured, registered. The perch has to be standardised for this purpose, made from steel, furnished by an installation to regulate the weight. This is not only a question of precision. The production of sports results is a mirror of a new societal pattern, of an emphasis on industrial productivity. At this point, the driving force of transformation in games as well as in society can be identified.

Under the guise of this new productivity, so *interpersonal relations* have also changed. The togetherness relating to gender, age, skill and strength has been dissolved by new classifications, meaning that participants become assigned to gender classes, age classes and weight classes. This has constituted a new form of 'equality', but at the same time has broken up and scattered the games' social connections. A new type of patriarchal–masculine dominance has also tended to become established in sport, while the dance, with its rhythmic vibration of togetherness, has disappeared and been set outside sport in a strict sense. That Brittany has kept the *fest-noz* (the night dance festival) has been a lucky 'anachronism'.

Associated with the new productivity of sport have arisen new values and mythologies, the most well-known being the Olympic idea (or ideal) of Pierre de Coubertin. The complex pedagogics and para-religiosity of the founder have, however, gradually been reduced to rather one-dimensional principles such as 'faster, higher, stronger'. The institutional equivalent of the ideas of record increasing, achievement and competition is the hierarchical system of tournaments and championships. Systems of this kind are organised and controlled by national and international sports federations and their bureaucracies, although in recent years commercial organisations, science and television have had an increasing influence on the institutional framework of the games.

SPORTISATION, PEDAGOGISATION, FOLKLORISATION

Sportisation has, however, been but one form taken by the modernisation of games (Figure 8.1). During modernity there has always existed another model of sport, and physical activity surviving alongside achievement sport. This has been variously termed gymnastics, sport for all, fitness sport or welfare sport. In this model, bodily movement is instrumental and functionalised: fitted for health and hygiene, for pedagogical adjustment, for social integration. The pedagogisation of Inuit games in Alaska was quoted earlier as an example. In such a situation games are loosened from their cultural context but in a different way: they are treated as a pedagogical instrument. As shown by the Indonesian fighting art of *pencak silat*, the game involved can also be developed towards a sort of militarism. A third model of modernisation is that of folklore, which means a third form of isolating the game or dance from its social background, but here in the claimed interest of 'preservation' and representation. The movement or activity is used as a sort of theatre: it presents national or regional identity, often in a bizarre manner, appealing to the fancy of tourists or state guests. Its spatial–institutional equivalent is the museum, and this model too was mentioned above as a risk in the development of North American Inuit games.

144

	Traditional games	Modern sport	'Postmodernity'
1. Time	Festivity Rhythmic repetition	Leisure/work time Futurist maximisation	Sport of the unemployed Experience of stress
2. Space	Integration Local identity	Specialisation Standardisation Isolation	Jogging space 'Community sport' Crisis of 'container architecture'
3. Energy	Laughter Music Ritual 'spirituality'	Disappearance of laughter Disappearance of music Dynamic of speed	New carnivalism Re-musicalisation New religious body cultures
4. Interpersonal relations	Patriarchy/ matriarchy? Togetherness of ages Locality/ internationality	Male dominance Classes of gender age, weight National identification/ results	'Sport across the limits' Family sport Social sport Multiplicity of cultural identities
5. Objectives	'The strong man' 'Glory'	Production of results Quantification of records Achievement	Crisis of educational quantification Media circus
6. Values and ideas	Christian/pagan? Traditionalism	Faster, higher, stronger c-g-s rationalisation Olympic ideology	Criticism of sport New Age ideology of sport 'Health' sport
7. Institutions	Local self- determination Age-groups?	Hierarchical system Bureaucratic control	Non-organised sport Commercial sport Alternative body culture

Figure 8.1 Premodern, modern and postmodern forms of games, sports and body
cultures

The question is, however, whether these forms of modernisation are today the only alternatives. Let us recall the remark of Andrzej Wohl about the new games: they can be regarded as reminders of medieval social structures, not as nostalgia but as a vision for the future. Does the new attention given to traditional games in the 1980s denote a new societal situation, one in which old games are not just a restoration of something anachronistic but a new experiment attaching to new conditions of life? Does the breakdown of the East European sports system signify a crisis of sportive modernisation, as well as of modern sportisation?

A POST-SPORTIST SOUND?

The *time* of modern sport, the wage worker's splitting-up of his or her day between leisure time and working time, is no longer relevant for the majority of the population. New social classes arise with new time patterns, and not only with those of the unemployed. At the same time, the futurist dynamic of sportive time seems to lose its fascination. What has been fascinating before, to raise the standard, is now turning to stress. Doping appears as a material expression of this stress situation. With respect to *space*, in the field of sports architecture and spatial planning we witness a growing criticism of 'container architecture' (Eichberg, 1988). Modern swimming facilities (water slides, irregular shapes) emphasise fun and *bodily* experiences rather than training or competition, and here commercial as well as alternative building practices are trying to make 'new waves'. Joggers, skateboarders, roller-skaters and carnivalistic festivals are re-conquering the road as a space of body culture and games. A new social politics is striving to use sports as a (central) factor in community reconstruction, in social network politics and in the securing of local identity. With carnivalistic traits in recent sport, the culture of laughter is reappearing, and we also witness a new musicalisation of sports: the unleashing of new *energies*.

Physical education in schools has been the first to reduce the production of quantitative results, to downplay the tyranny of *objective* assessments of who is 'the best'. Television is favouring top achievement sport in many ways, but on the other hand it is contributing to the reverse, shifting the focus towards the spectacular, towards sport as theatre, as circus. Social sport, family sport and sport across age limits are establishing new patterns of *inter-personal relations*. The 'Gallo popular games' at Monterfil since 1977 is an example of how to reconstruct social relations from the practice of traditional games, albeit in a new context. It is not by accident that the communicative aspects of traditional games are rediscovered in this situation (Møller, 1984).

These 'postmodern' and 'post-sportist' tendencies are not at all unequivocal, since there are also opposite developments. But the configuration of sportive modernity *is* in question. Studies in economic transformations since the 1960s show, moreover, that sport is not an isolated field in this connection. Economic and technological development is pushing in the direction of small-scale production and regionalisation, properties that – under the dominance of industrial modernity – had once been regarded as 'archaic' (Toffler, 1980; Piore and Sabel, 1984). If this is correct, then the tendencies in Western Europe towards autonomy on smaller regional levels are not accidental and not without a larger societal perspective: they are feeding into the political field as well as into the realms of games and body culture. Some 42 million Europeans are without or lie outside states speaking their own language: this is a large number; but they are only a part of the challenge, and not only on the level of linguistic self-determination. The same might be true for the revolution in Eastern Europe (Riordan, 1990), the comparison showing that the collapse of centralist state systems and attendant modern sports systems is not only a question restricted to the eastern hemisphere. In this process, the nationalisation of sports and games, which could be recognised from the late eighteenth century onwards, is lifted on to a new level. It becomes dramatically re-actualised and – this is its paradox – at the same time transformed and dissolved. National identification with sports results, national gymnastics, military discipline and national standardised folklore has come to a crisis situation, in parallel with that of the existing 'national state' (or nation-state). Popular cultures are finding their expressions in other body cultural forms, since cultural identity forms itself on the basis of a multiplicity of configurations. The drum dance is not a modern sport. If it is nevertheless significant for the late twentieth century, could it be a 'postmodern' sound?

BIBLIOGRAPHY

Audibert, A. *La matriarcat breton* (Paris, 1984)

Bale, J. 'Sport and national identity. A geographical view', *British Journal of Sport History*, 1986

Bernett, H. *Der jüdische Sport im nationalsozialistischen Deutschland, 1933–1938* (Schorndorf, 1978)

Blecking, D. *Die Geschichte der nationalpolnischen Turnorganisation 'Sokol' im Deutschen Reich 1884–1939* (Münster, 1987)

Bogusz, J. (ed.) *Polish Youth* (Warsaw, 1988)

Carret, C. *et al. Les jeux de palets* (Rennes, 1983)

Chatschkowanjan, G. 'Athleten auf der Matte', *Sowjetunion heute*, 1972

Daudry, P. *Documenti di sport popolare* (Aosta, 1981)

Daudry, P. (ed.) *Jeux et jouets de la tradition valdotaine* (Aosta, 1990)

Denez, P. *Mouvement culturel breton et développement économique et social* (Quimper, 1988)

Düding, D. *Organisierter gesellschaftlicher Nationalismus in Deutschland, 1808–1947* (Munich, 1984)

Dufaitre, A. *Traditional Games* (Strasbourg, 1989)

Dunning, E. *The Roots of Football Hooliganism* (London, 1988)

Eichberg, H. *Der Weg des Sports in die industrielle Zivilisation* (Baden-Baden, 1973)

Eichberg, H. *Leistung, Spannung, Geschwindigkeit* (Stuttgart, 1978)

Eichberg, H. *Museet der danser* (Gerlev, 1985)

Eichberg, H. *Die Veränderung des Sports ist gesellschaftlich* (Münster, 1986)

Eichberg, H. 'Dansens energi', *Centring*, 1987

Eichberg, H. *Leistungsräume. Sport als Umweltproblem* (Münster, 1988)

Eichberg, H. 'Trommeltanz der Inuit', in Eichberg, H. and Hansen, J. (eds) *Körperkulturen und Identität* (Münster, 1989)

Eichberg, H. and Hansen, J. (eds) *Körperkulturen und Identität* (Münster, 1989)

Einarsson, T. *Glima. The Icelandic Wrestling* (Reykjavik, 1988)

Floc'h, M. and Peru, P. *C'hoariou Breizh. Jeux traditionnels de Bretagne* (Rennes, 1987)

Franco, R. *Juegos y deportes Vascos* (San Sebastian, 1978)

Frey, R. and Allen, M. 'Alaskan native games. A cross cultural addition to the physical education curriculum', *Journal of Physical Education, Recreation and Dance*, 1989

Frogner, E. 'On ethnic sport among Turkish migrants in the Federal Republic of Germany', *International Review for the Sociology of Sport*, 1985

Godio, R. (ed.) *Sport tipici regionali* (Saint-Vincent, 1985)

Guttmann, A. *From Ritual to Record* (New York, 1978)

Guttmann, A. *A Whole New Ball Game* (Chapel Hill N.C., 1988)

Hargreaves, J. 'Urban dance styles and self-identities', in Eichberg, H. and Hansen, J. (eds) *Körperkulturen und Identität* (Münster, 1989)

Hass, D. 'Die besten Köpfe gehn weg', *Der Spiegel,* 1989

Hein, C. *Der fremde Freund* (Berlin, 1982)

Hellspong, M. 'Traditional sports on the island of Gotland', *Scandinavian Journal of Sport Sciences,* 1989

Hörbe, P. *Sport und irische Geschichte,* diss. (Cologne, 1983).

Ipellie, A. 'A look back at Northern Games '78', *Inuit Today,* 1979

Jaouen, G. *et al. Ar gouren* (Rennes, 1985)

Johnston, T. 'Alaska Eskimo music is revitalized', *Dance Research Journal,* 1978

Korsgaard, O. *Kampen om kroppen* (Copenhagen, 1982)

La Borderie, A. 'Luttes bretonnes au 16 siècle', *Revue de Bretagne et de Vendée,* 1988

Loest, E. *Es geht seinen Gang oder Mühen in unserer Ebene* (Stuttgart, 1978)

Lynge, B. *Rytmisk musik i Grønland* (Aarhus, 1981)

Mangan, A. *The Games Ethic and Imperialism* (London, 1986)

Moëlo S. and Le Bihan, J. *Kergohann hag ar Vellad* (Loudéac, 1986)

Møller, J. 'Sports and old village games in Denmark', *Canadian Journal of History of Sport,* 1984

Møller, J. *Gamle idrætslege i Danmark* (Kastrup, 1990)

Natschkebiya, K. *Georgian Equestrian Folk Games* (Tbilisi, 1964)

Pedersen, K. and Rafoss, K. 'Sports in Finmark and Saami Districts', *Scandinavian Journal of Sport Science,* 1989

Peitersen, B. and Kristensen, B. *An Empirical Survey of the Danish Roligans during the European Championships '88* (Copenhagen, 1988)

Peru, F. 'Le lever de la perche en Trégor', *Ar Men,* 1986

Piore, M. and Sabel, C. *The Second Industrial Divide* (New York, 1984)

Redmond, G. *The Sporting Scots of 19th Century Canada* (Toronto, 1982)

Renson, R. and Smulders, H. 'Research methods and development of the Flemish Folk Games File', *International Review for the Sociology of Sport,* 1981

Riordan, J. 'Folk games and fake games in Soviet times: the case of Gorodki and Lapta', *Stadion,* 1986/7

Riordan, J. (ed.) *Soviet Youth Culture* (London, 1989)

Riordan, J. 'Playing to new rules: Soviet sport and perestroika', *Soviet Studies,* 1990

Sandblad, H. *Olympia och Valhalla* (Stockholm, 1985)

Schröder, W. *Burschenturner im Kampf um Einheit und Freiheit* (Berlin, 1967)

Soares, A. (ed.) *Jogos tradicionales do Ribatejo* (Santarém, 1980)

Stejskal, M. *Folklig idrott* (Turku, 1954)

Toffler, A. *The Third Wave* (London, 1980)

Williams, J. *et al. Hooligans Abroad* (London, 1985)

Wohl, A. *The Scientific Study of Physical Education and Sport* (Gerlev, 1989)

Zurcher, L. and Meadow, A. 'On bullfights and baseball', *International Journal of Comparative Sociology,* 1967

9

THE SOCIETAL CONSTRUCTION OF TIME AND SPACE AS SOCIOLOGY'S WAY HOME TO PHILOSOPHY
Sport as paradigm

What happened, fundamentally, in movement culture and society when sport invaded everyday life and made it 'modern'?

In 1893, the first comprehensive Danish book of sports was published (Hansen, 1893). It presented among other things some older forms of foot-race that never became real modern sports, but which hypothetically at least could have become so. These included the sack race and a race with water buckets on the head. Why did they have no chance, and why cannot one imagine today a Danish national tournament for water-bucket racing or an International Federation for Sack-Racing? (Figure 9.1). Questions like these are highly relevant for any reflection on what sport has become in the present. These non-sports included, or could easily have included, all of what the ideologists (and some sociologists) of sport have declared to be essential for the conduct of 'real' sport: rules, competition, fair play. But these were evidently not the decisive criteria determining the historical transformation of certain bodily activities, but not others, into recognised sports. Where can these criteria be found instead? Alongside pictures of water-bucket and sack-racing, the Danish sports book presented another showing the new sportive practice of athletic running competition (Figure 9.2). This picture tells, through its comparison with the non-sport races, what really was becoming specific to sport. It tells about space and time. It shows the modelling of space and time in sport, and suggests – indirectly – how space and time were to be constituted in the mainstream of societal common sense and social practice.

Time in the conventional (sports) sense has now become what can and should be measured by the stop-watch as precisely as possible, thus making human achievement objective. More precisely:

1. Time is *one-dimensional* and absolute; it is not poly-dimensional or relative.
2. Time has a *direction* just as running is directed towards a goal; it is irreversible.

Figure 9.1 Sack racing

Figure 9.2 Track racing

3. Because of its imaginary scalar–directional pattern, time can be *measured*, divided, compared and classified (just like money, leading to the phrase 'time is money'); the time of sports is the time of the stop-watch.
4. Quantification of time means that time can be fixed in the form of a *result*, a produced figure in seconds (or, more generally, sports can be captured in terms of c-g-s, or centimetres-grams-seconds, and points); this means that results can be compared and *records* established.
5. But why quantify, measure, produce and compare? Because time here is a field of *speed* and *acceleration*, the records will always rise towards an open horizon – quicker, quicker, quicker (or, more generally, *citius, altius, fortius*)

150

– and there is hence a growth in sport as in time, and never a falling of the record (Eichberg, 1978, 1989a; Hopf, 1981; Sachs, 1984; Penz, 1987, 1990).

Space is related to this configuration of time in different ways. It is the framework that can and should be measured, structured and standardised, thereby fitting the functional needs of time-directed achievement production. And in effect time is made visible by space, using visualisation techniques like the watch or the scale or verbalisations like 'short' or 'long time'. More precisely:

1. Space as seen in the sport picture is dominated by the *straight line*; it is *panoptical*, creating the impression of a total survey.
2. By *borderlines*, space is divided into different lanes or zones for the runners so that they cannot get bodily into contact (as they had done in many premodern running events), and this parcellation is doubled by a segmentation of the landscape in many different sports: the space of the foot race cannot be used for football, jumping or tennis, and in the further evolution of these sports it cannot even be used for horse-, cycle- or motor-racing; sports space is tending to become *monofunctional*.
3. The ideal form of sports space as a three-dimensional architecture is the *container*, a box of right angles; the monofunctionality of this sports container means that different sports need different specialised boxes.
4. The 'function' of sports space is to further the production of results; and for this purpose the space must be *standardised* according to universal norms that are determined by various levels and agencies of sports bureaucracy; this development makes the space of sport independent of concrete places, rendering it unspecific, placeless.
5. Besides its geometrical container aspect, however, the space of sport also contains dynamic elements in the shape of spatial *expansion* whereby many sports disciplines have sought to extend their 'straight-lining' tendencies into a control over, or denial of, the irregularities of outdoors space; and repeatedly and dialectically, during the process of sportification outdoor movements have themselves become domesticated and caught in between walls (Eichberg, 1988, 1993; Bale 1989, 1993; Puig and Ingham, 1993: see also Chapter 3).

MODERNITY AND RATIONALITY IN THE RACE

The pattern of straight-lined space and one-dimensional time has not been the only possibility in the history of sports, but contrasting configurations demonstrate – by their ephemeral or marginal existence in modern movement culture – what really was becoming hegemonic. In older folk practices, races with handicaps were particularly popular. As we could see from 1893, experts in this early phase of sports may still have expected the sportification

151

of such events as the sack race, water-bucket race or three-legged race, but the configuration of these competitions did not fit together with the straight rationality of time–space–results. In contrast, these games celebrated the stumbling, the curved and crooked lines, the grotesque body, the laughter. These events soon disappeared from serious sport and no three-legged race is now found in the Olympic Games.

Another side-track of running was the forest race. Here too the experience of space and time formed a contrast to the standardised track: by the variety and surprise of wood and trees, by curved ways and obstacles, by the experience of smell and season, by changes in landscape and nature. Having gained some popularity, especially in the years around 1900, the wood race still did not end up being a characteristic modern sport (even if it did comprise the starting point for the later sport of orienteering).

A further aspect of sportised running gained more importance in the framework of gymnastics: the disciplining of running techniques. By strict exercise forms, changing the running positions after the command of the gymnastic leader and following the rectangular space of the gymnasium hall, this activity expressed another side of Western rationality: less product-oriented, but as a form of abstract body discipline. By running, the straight body was modelled, trained and educated in a container hall environment, and running was here a means of disciplining the 'right' body poise, forming the straight backbone. The comparison of these different types of running and racing shows that sport is indeed not 'natural', but a very specific staging of space and time. And it also shows that neither are space and time themselves 'natural' and universal. The potentially multiple but nevertheless hegemonic structure of modern racing and running suggested that there is no 'natural balance' or equilibrium between space and time. The natural-scientific term of 'dimensions' is too abstract to describe the concrete societal modelling of time–space configurations. Time and space do not simply stand side by side: instead, modern sport seems to establish a domination of time over space. This becomes clear by comparing it with the noble exercises of the seventeenth and eighteenth centuries – courteous dance, rapier fencing, figure riding, horse vaulting and court tennis – which had established a dominance of spatial order over temporal dynamic, a social geometry of the body. Modern sport, in contrast, does not mean to structure a geometrical world by well-mannered, decent, rhythmic motions; rather, space has tended to 'disappear' in the process of sportification, becoming an effaced panorama, a blurred horizon, a standardised channel for the streamlined body projectile. Space duly furnishes the uniform conditions and framework for the sportive time-dynamic production of results. There are far fewer sport records in centimetres compared with records in seconds, and even these few 'spatial' records (as in jumps or throws) are bricks of a dynamic time configuration, by their tendency of transgressing the established horizon in the pursuit of an open future. The hierarchy of modern movement culture can be formulated as: *producing results – accelerating time – standardising space* (or expanding it).

The same is true outside the hegemonic race-and-stop-watch model, in football for instance. Premodern folk ball-games had often occupied, integrated and discovered the whole space and landscape between two (or more) villages or quarters. In modern soccer, however, the relevant sports space has shrunk towards a standardised plastic field serving the production of 'goal results' and the time orientation that modernity calls 'tension'. (The 'underground' dimension of football space, the 'dance' of the player and his or her 'acrobatics', is another story.)

Sport is thus a visible, living metaphor of modernity: making the world measurable, making the human being productive and 'developing' space under the dominance of 'dynamic' time. The one and dominating rationality finds its expression in pure form through the time–space patterns of sport. But is this still the case? Recent innovations in the field of sport are questioning the classical modern image, and when examining 'the rationality' of time in the overall panorama of today's movement culture one can find a remarkably different picture.

STRESS, NEW SLOWNESS AND SITUATIONAL TIME

The classical modern fascination of directional time, speed, acceleration and tension has certainly not disappeared at the present time. It holds its place, especially in the field of spectator sports, and notably in the new media circus existing side by side with crashing car pursuits and the thrilling showdown of entertainment films. In popular mass activity, however, something like jogging may signal a fundamental change of movement culture: raising the possibility of running not as a creation of productive tension with the stop-watch at the horizon but, on the contrary, as a technique of de-stressing life. Stress casts a shadow over the modern world, appearing as the dark double of futurist time enthusiasm: no time, no time, no time! This double shows up obviously in competitive sport, which has become a major field of stress production, side by side with industrial work and school education (Nitsch, 1981). From the fascination of the Olympic *citius* (quicker, quicker, quicker) to stress, from time dynamic to its reverse, to illness: this has been the way of sports. In this respect, jogging and other similar mass activities might rather be thought of as anti-sports: they are creating islands of timelessness in modern life.

Western movement culture since the 1960s has been marked by the expansion of East Asian exercises, by yoga, t'ai chi chuan and older techniques of meditation and concentration. These exercises are following older movement patterns to create a new slowness, contrasting sharply with the configuration of modern speed and acceleration. Yet this new slowness has not remained solely a feature of an alternative movement culture, but has

153

invaded the mainstream too, as meditation and 'inner sports' become means of training for better results and raising the standards on the running track, on the football pitch and elsewhere.

When the so-called 'New Games' entered the world of sports around 1970, they did not only represent (following their original intention) another 'political' attitude: one of togetherness instead of winning, fantasy instead of standardisation. In fact they bore a new time pattern as well: people were playing the game here and now without reference to earlier records. Together with the importance of results, the time curve of tension disappeared and was replaced by a sort of situational time pattern. The New Games stepped forth as 'never-ever games': games that could certainly be repeated, but could also be played here and now and never again. Some New Games could become sportified – as was the case with frisbee, skateboard and surfing – but this meant transferring them into the time expectations of modern sport. A physical movement in and of itself is evidently not 'sport' (nor 'game', nor 'dance', nor 'work'), but the differing social time pattern is decisive. Under this aspect, the 'zapping' of time by the New Games is relevant, and not least because it reappeared in the boom of 'traditional games' during the 1980s.

In the years from 1770 to 1820, the waltz had marked a revolutionary body process, both as a new social configuration – entailing the 'revolution of the couple' (Hess, 1989) – and as a new temporal configuration built upon speed and a frenzy of whirling and dynamic forward movement. This is no longer the only, nor even the dominant, pattern of social dance in the late twentieth century, especially among young people. Rock culture, a new type of 'somatic' or 'belly music', and related forms of dance and body staging took over in the 1950s/1960s and in so doing finished with the established pattern. Since 'twisting' at the beginning of the 1960s, the couple structure began to dissolve, and the forward movement came to a standstill to be replaced by polycentric body movements (Eichstedt and Polster, 1985; Klein, 1992). The time structure of rock as 'somatic music' was born by the rhythmic sound of the drum, and this renaissance of the drum represented a fundamental breakthrough in Western music, where the drum had been banned since the Middle Ages (with military march music as a single exception) and where the Afro-American jazz of the 1920s constituted a forerunner of the new 'exoticisation by drumming'.

FESTIVITY, BIOGRAPHY AND SOCIAL LIFE-FORMS: IDENTITY IN TIME

The new movement culture arising since the 1970s and 1980s is often organised around both festivals, as in the case of rock culture, and other arrangements of festive character. Festivity had been the social–temporal framework of premodern games, pastimes and dances, but in the process of modernisation and sportification the principle of festivity became replaced by one of 'discipline', the specialised branches of single activities. From the multi-dimensional carnival, *kermis* or tournament, this development led to one-dimensional federations of boxing, basketball and sailing. The principle of discipline meant a time pattern of continuity and gradual perfection, of repetition and increase – effective indeed – integrated into the modern temporal parcellation of work versus leisure. In consequence, any re-festivisation of sports today means not only changes at the level of contents, a recarnivalisation and re-musicalisation of sports, but also a change of time patterns from linear growth to rhythmic repetition and punctual, situational discontinuity: thereby giving life to the *event*.

Sport is also changing 'age'. What once had been a feature of youth culture is nowadays (as figures on participation in organised sport clearly show) mostly a preoccupation of both children and 'grown-ups', and even of aged people. This is far from being only a demographic process, for it is cultural and concerns the wider change of movement patterns. The 'growing older' of sports can, for instance, be related to the new slowness in a manner different to how youth racing might be so related. But more than this, since a new relation between sport and life history seems to be on the way, one pivoting between movement and biography (Eichberg, 1994). Sport as youth activity could readily be placed on the linear upward curve of perfection, whether called career, growth or progress, which has been integral to the directional imaginings of modern life (and as haunted by the parallel icon-ography of old age as decay or degeneration). Yet sport as a part of the world of elderly people is now entering, albeit more ambiguously, into a universe of biographical discontinuity, of ups-and-downs, of health and illness, of marriage and divorce, of structural change and shifts in identity. The modern self-evidence of combining the 'young and sporty' (perhaps in the guise of some Ministry of Youth and Sport Affairs) is hence becoming fundamentally anachronistic.

Sport has taken part in the modern split-up of life and time, in the parcel-lation of work and leisure. The place of sport has, as a result, been con-structed as set within the world outside work. Thinking about sport meant thinking about the quality of leisure time, and sports research tends to be couched as part of leisure studies. Recent studies in social life-forms have shown that this does not at all represent the totality of modern everyday life, however, and that – moreover – it never did (Rahbek Christensen, 1987). The dichotomous time pattern of leisure (and sport) versus work may characterise one – admittedly large and representative, but rarely politically hegemonic – class in industrial modernity, that of the wage workers. This class grouping works to satisfy need, and balances this strenuous work with a realm of freedom in leisure time. It also tends to secure its identity in this latter realm, and the implication is that sport can become at least as import-ant a source of identity as other factors in working-class life. Another and more powerful class has built up a similar dichothomy, although with inverse values. Thus career-oriented, mostly intellectual, pedagogical, administrative or managerial workers build up identity in their work time and through their work, while leisure – including sport – is mostly relegated to supporting this work-based identity, for instance as demonstrative consumption. A third class, encompassing the independent professions like farmers, craft-workers and small shop-owners, rarely exhibit any comparable parcellation of work and leisure, but rather integrate their family life and their identity into the operations of their enterprise. Two 'female' groupings do not live the split-up time pattern either: the housewife and the hinterland wife, as related to the wage worker and the career-oriented worker respectively.

Time – as shown by this analysis – is not an 'objective' scientific frame-work to be measured and classified, but is instead a relation that interacts with wider patterns of social identity. The place of sport in time is hence related to the question of sport and social identity, taking seriously sport in its connections to class, ethnicity and, last but not least, gender. A new, and at the same time not so very new, multiplicity of times becomes visible in the process, contesting the one-dimensionality of sportive stop-watch time as the hegemonic time pattern of modern rationality.

DISNEYFICATION AND LABYRINTH

The spatial configurations of sport have not remained unchanged during the process of temporal 'postmodernisation', as will now be discussed. At the same time as the rationality of sportive time was beginning to be experienced as stress, for instance, in the field of architecture a violent criticism began to be directed against the functionalism of building. This criticism was based on a new sensitivity to the spatiality of the monofunctional container, and found its form in a new expressivity of building permeating general archi-tecture as well as sports facilities. The alternative 'movement house' appeared as a Danish innovation, while Hungarian 'organic' architects experimented with both ex-centric spaces and halls for dance, game and sports (Garborjáni and Dvorsky, 1991; Eichberg, 1993). Strictly commercial interests were no less sensible to discovering this new trend, making profit from it by creating new, exotic bathing landscapes and non-standardised adventure parks for leisure sport, and so the panoptical sport container suddenly found itself caught between the scissors of an alternative expressivity on the one hand and a commercial 'Disneyfication' on the other.

Maybe this new building activity and the attendant critical architectural sensibility has been based on (or at least related to) a more fundamental innovation at the level of body movement. Jogging, skateboard activities and roller-skating have penetrated and occupied the territory of cities, on the surface as well as in the subway, while climbing sport is conquering urban environments in the third dimension, whether using the interior of old factories or the outside façades of office towers. It is still difficult to express this new experience of spatial adventure in words. (The sportification of these activities, as in sponsored skateboard competitions with tournament rules and specialised indoor facilities, is still part of 'the old story'.)

Several activities of the new type involved here follow a labyrinthine pattern of space and motion, in which the actor is winding on curved ways like the graffiti on the walls showing non-panoptical iconographies (nonsense figurations, but not at all without significance). At the same time, a new interest in the classical labyrinth has arisen, documented in innumerable labyrinth publications and speculations, in crowded expositions and in new labyrinth or maze buildings. Contrasting the straight line of racing, the labyrinth seems to tell a bodily history that is related to 'postmodern' curiosity, and to the mazes of computer games that are children's entry and socialisation into information society (Eichberg, 1989b).

The curved labyrinthine ways are also leading the jogger outward into an environment of woods and green landscapes. A lot of new movements and outdoor activities have been forming a 'green wave' since the early 1970s, whether woodcraft movements after native American or Nordic models or manager courses in the wilderness, or whether adventure survival games in the forests or ecological 'soft tourism'. Maybe this wave, with all its contradictions, is the bodily substratum of what becomes an ecological awareness on the plane of mind soon afterwards (Eichberg and Jespersen, 1986). Again, the sportification of some of these activities has been related to a colonisation of space, thus representing the 'old story' of modernity, but this is only one side of the coin. The other side is the 'inner ecology' of this green wave (Moegling, 1988), as expressed in a green reappropriation of the urban centres by marathons, sport festivities and individual jogging, and also through a spiritual approach to body ecology practised through 'green meditation' or 'sitting outside', a Nordic experience in night and nature.

NEW REGIONALITY AND NO POINT OF SURVEY: IDENTITY IN SPACE

The festival of games has another spatial order (or chaos) to that of the typical modern sports tournament. The Olympic Games as a pseudo-festivity and (in reality) a super-tournament demonstrate an extreme state of spatial colonisation, covering the towns with monofunctional concrete containers and fortifications, all connected together by more and more extensive transportation systems. This hinders the participants, whether actor or spectator, in ever getting a view of the whole of the games. The opening and closing ceremonies are only the alibi, a showtime compensating for the lack of genuine festive connection. The games festival, in contrast, temporarily appropriates the 'normal' space of life, the village green or the urban centre, barring the motor traffic and using the place counter to its usual functions.

This becomes visible in both New Games festivals and traditional games fairs, in rock festivals as well as in city marathons, or in something like the annual Games Day of the Danish *folkelig* sport. Where modern sport – by its principle of discipline and specialisation – tends to homogenise the national territory by covering it with a growing and increasingly refined screen of expert groups, specialised national organisations and monofunctional facilities, the space of the festival is punctual and local, combining different kinds of movements within one situational event here and now. Where the space of modern sport has become the territorial screen, the space of the games and festivals is the *place*, and the place, with its unique quality and ecology, is therefore the configurational correlate of what in the temporal dimension is the *event*.

The configurations of both the specialising territoriality and the situational place respectively create or signify differences at the level of identity. Modern sport has been an agent of nation-building, knotting together individuals as experts on the territorial, state, international and universal levels, and, with the help of homogenising rules, sport can give athletes a simultaneously equalised and hierarchical position relating to territorial identity: as provincial top athlete or as national champion. In contrast, the festive event, by being tied to a situational place rather than to a territory, forms identity in another way, which is still difficult to express in words. (We know from pre-industrial game cultures, though, that the carnival was not a homogenising factor or one bound up in constructing territorial state identity.) In the games or rock festival, there is no pyramid of spatially homogenised competitions resulting in a champion who finally represents the whole but, instead, what is cultivated is the memory of a unique place and event and its emotional intensity. So what appears today as a new regionality of games – in Caucasian Georgia, in Brittany, in the Basque country, in the Valley of Aosta, in Tatarstan – or even as a new nationalism, expressing itself through festivals of folk dances, popular sports and traditional games, may arguably point more in a new transmodern and transterritorial direction.

What makes it difficult to define the spatial features (as well as the temporal features) of the new movement culture might be part of the phenomenon under analysis itself: namely, the non-panoptical structure integral to this new spatiality of sport. Space and identity (like time and identity) here means many spaces (or times) and many different spatialities (or temporalities), all differentiated along the axes of gender identity, regional culture, subcultural identity and historical change. While the pyramid – the ideal image of the modern sports system – is spatially oriented towards one central point of survey, which is the panoptical 'top', the labyrinth does not know any such panopticism. At no place in the classical labyrinth or in the maze can one gain visual control over the situation, and even in the centre of the labyrinth the opportunity for survey is lacking. The labyrinth seems iconographically to tell a counter-history, one running against the hierarchical totalitarism of the central perspective. What the nostalgic defender of modernity and its rationality complains about as the actual 'Neue Unübersichtlichkeit', the new lack of survey (Habermas, 1985), is evidently not only a philosophical idea but also a bodily practice and a spatial experience. Maybe it was no accident that the epoch of these labyrinthine movements gave rise to the discovery of fractals, thereby giving another way of picturing the complexity of social practice.

It is difficult to determine – and impossible to quantify – the significance of these configurational changes. Are they just a curl at the surface of sport and society, or does sport – by in effect producing a 'non-sportive sport' (Dietrich and Heinemann, 1989) – signify a key change in the deep structure of modern Western sociality? The geometrical dances and exercises of early modern Europe, together with the epochal innovation of central perspective, had indeed prefigured the rise of the Western territorial state with its social-disciplining absolutism. The forward movements in the dances and sports of the 'Age of Revolutions' likewise paved the way towards modern industrial society. And so, do the new slowness and the situational time, the new spaces of place and labyrinth, support the scenario of transmodernity? Whatever the answer, this study of space and time shows that they are more than merely co-ordinates of a quantitative rationality. Rather, they form configurational patterns that – on a level of superstructure embracing ideas and discourses – take shape as elements of societal mythology, and in this respect sport is much more than just sport.

PROGRESS, CAREER AND HIERARCHY: THE RELATIVITY OF SOCIETAL MYTHOLOGIES

Seen from a distance that is affected by the recent innovations, what do the configurations of traditional modern sport tell us about the dominant mythologies in industrial society? And, what is the social language of the accompanying shift in body culture?

'Progress' has been a term of fundamental significance in racing sport as well in societal thinking, linking into terms such as 'mobility' and 'acceleration', 'achievement' and 'growth' (Koselleck, 1975; Oettermann, 1984). Together, they have formed a pattern that could readily be understood socially because of its base in bodily – spatial, temporal – experience. But this could only work in a society esteeming highly the race and the stopwatch in sport. Professional runners of the 1830s set the word 'progress' on their flag, and at about the same time – and later – 'progress' appeared as a political direction on both (red) flags and in (left-wing) newspaper titles (*Avanti, Vorwärts, Fremad, Fortschritt, Progress*). The contents of such 'progress' would always be controversial and disputed, but even the 'conservative' or 'reactionary' (right-wing) positions argued from inside the same configuration, whether they demanded to 'slow down' the race, to return it to the

159

starting point or affirmed that their schemes represented 'the real progress'. It was not before the 1970s, and especially after 'The limits to growth' text published by the Club of Rome, that relevant sections of Western societies became aware of the problems implicit in this terminology (and iconography). Maybe this was more than a purely intellectual discovery, also entailing a shift of configurations based on new experiences of race and non-race in sport and dance, of the situational, the labyrinthine and the new slowness. Since that time, societal thinking cannot any longer use terms like 'progress' without irony, without excuses or without reflection on its strange relativity: 'progress' is now fiction.

'Career' had in early modern times been a word to designate the quickest galloping movement of a racehorse. With the rise of pedestrianism as an early sport, the term was transferred to human running and to the race-track, there describing the one-dimensionality of motion and its futurist directionality: running in the one given direction to secure the prized result. 'Career' became on the individual level what 'progress' meant on the collective one, and it duly became a fundamental image for pedagogical thinking as well as for the organisation of professional life according to modernity. There always remained some differences between distinct social classes and life-forms, however, with some of them standing further from the possibility of having a 'career' (farmers, housewives and craft-workers, for instance) while some of them standing nearer to it (the career-oriented professions). But, on the whole, the image seems to have survived intact, at least until the 1960s and 1970s, when unemployment became permanent and structural. Now, youth cultures have developed notions of 'no future' and 'you have no chance – use it', expressing their concerns in the spatial dimension through phrases such as 'under the pavement, there is the beach'. Sociologists have begun to discuss 'the end of the labour society', and – while this has been mostly described by fashionable theories as simply a 'change of values', thus remaining on the surface of social life – it may (once again) be indicative of something more than just a fashion and an intellectual wave. Indeed, it may even bear witness to more than just an automaton's response to changing economic 'facts', pointing instead to a social–bodily awareness of the relativity (and fictionality) of the modern 'career'.

The 'pyramid' has always been the ideal pattern for organising and visualising modern sport in both East and West. With its one-dimensional orientation 'up and down', along with its classificatory structure and its panoptical clearness of control, it has formed a frame for understanding individual 'careers' as well as collective 'progress'. The small 'pyramid' for the three winners receiving their medals beside the race-track is configurationally related to the universal bureaucratic hierarchy of Olympic sport. The ideal hierarchies of both Fascist parties and capitalist enterprises have undoubtedly followed the same scheme (although, as some in-depth studies have shown, not always in the real structures of power and everyday life underpinning such edifices), but even the most egalitarian and allegedly class-less systems of modernity have followed suit in representing themselves by 'pyramids'. The former Soviet Union and GDR did not hesitate to use – quite officially – the 'pyramid' and even the terminology of 'classes' in order to describe their model of sport. This was contradictory to aspects of communist ideology, but logical and consistent under the aspect of modernity. It seems that the breakdown of political–bureaucratic hierarchies in Eastern Europe around 1989 was prefigured by a crisis of this 'pyramid' model in sports, which can be documented since the 1970s and 1980s. If this observation is right, and if the connection between the panoptical hierarchy and modernity can be affirmed (following Lewis Mumford and Michel Foucault), then the revolutions of 1989 had not only an anti-totalitarian significance but also a transmodern (and trans-sportive) one. The breakdown of 'pyramids' in sport and society would in this case not only (or not primarily) signify the wish for new 'pyramids' – for fresh patterns of Western sport along-side capitalist trusts and corporations – but rather a strive for 'fractal' and 'labyrinthine' models of sociality.

Parallel to and connected with 'progress', 'career' and 'pyramid', other societal mythologies could be analysed with reference to their sports-related aspects and their significance for (or basis in) body cultural practice. The panopticon as a means of survey in exercise and society formed, together with a parcellation of (sports) activities claimed as socially 'functional', the modern archipelago of control and disciplination (Foucault, 1979). Achievement in sport and society, crystallised in imaginary products like results or records in centimetres, grams, seconds or points, manifested the rationality of industrial productivism. The syndrome of speed and acceleration – 'dromocracy' – reduced the environment to a bypassing panorama, thereby figuring the inner problems of ecology (Schivelbusch, 1977; Virilio, 1977, 1978; Sachs, 1984; Eichberg, 1987). Yet the phenomenology of happening and situationism (Dressen, 1991), the zapping time of media, the new labyrinths and the fractal structures of space and society, all show that the long-time established trends of modernity are no longer stable. And so we have to ask which mythologies, 'after modernity', will result from this destabilisation, and how they will be related to sport and bodily rituals.

BASE AND SUPERSTRUCTURE REVISITED

The connections between time–space configurations and mythological icon-ographies in sport and society can be interpreted on different levels. On the level of semantics, society can be understood as speaking by metaphors related to primary experiences in body and movement (O'Neill, 1985; Johnson, 1987). This claim directs attention towards a collective subconsciousness that is permanently producing, through its metaphorical operations, social knowledge: or, in other words, that is using bodily knowledges (such as that of the foot-race) as the basis for even the most abstract and scientific discourses (such as 'progress').

On another level, the connection can also be described, in its very explicit organised forms, as related to various ideologies and cults. Some of these explicit movements – especially ones related to sport and time – accumulated at the beginning of the twentieth century. Taylorism translated sportive features like quantification, competition and the stop-watch into industrial work, creating a new form of scientific human technology. Futurism glorified sport and the racing-car as symbols for a future promising a streamlined relation of the human being to time and progress, as contrasted with 'passatism': destroy the museums and replace them with sports grounds! In Italy, Fascism could use this imagery for its sportive programme of health, culture and youth, while Russian Futurism ran into the Soviet Proletkult combining mass sport, mass theatre, ergonomic laboratories and psycho-technical experimentation with an (anti-religious) religion of technology and also a cult of time measurement. Later, Soviet 'anthropomaximology' renewed some of these visions by declaring sport to be the base for a perfection of the human being on all of its physical, biological and psychical levels. Functionalism and the ideological movement of technocracy pushed similar ideas forward in the Western countries, although in practice functionalism was always less functional and more an aesthetic cult (Lichtenstein and Engler, 1993). However eccentric the explicit cult movements and mythological discourses might have been, though, they were undoubtedly highly significant. They expressed societal expectations at the level of superstructure, gaining their success from references to the more mundane bodily experiences in social life, including in particular those associated with sport.

But what is constructing what in society? Are the times and spaces of movement culture to be placed in the material 'base' or in the 'superstructure' (to use basic Marxist terms)? If the bodily experience and practice of time and space in everyday life – including work, sport, dance, games and festivities – is the base of societal processes and contradictions, then sport is not only an ideology present on the surface of symbolic culture nor an organised system and institution only featuring as part of society's superstructure. Instead, as a ritual of 'the social body', it is rooted deep in the material fundaments of society, figuring patterns of work and even prefiguring revolutionary changes. So the old question of what is base and what is superstructure in social life, and in societal change, is fully alive. How does movement culture (with sport as one specific paradigm) relate to implicit knowledge and how does it feed into explicit ideas and institutions (Polanyi, 1966)? And what is changing within the implicit level – perhaps being more visible in sports and dances than in other societal spheres – before the explicit consensus collapses?

And, last but not least, what is space and what is time? We started by analytically separating them, as is usual in the Western tradition of thinking.

But their changes seem to be interconnected, just as there is obviously an indissoluble time–space fusion in movement. Social time and social space are not only two, they are also one. Reflecting upon space and time through the experience of sport thus becomes a basis for critiquing the whole reification of space and time in Western thought. Indeed, through this configurational reflection upon time and space, sociology (with the unanticipated help of sports sociology) returns to where it had been at home in much earlier centuries: to philosophy; to a critical philosophy of life. Actual changes in time–space configurations demonstrate that time is more than the stop-watch and the streamlined body, and that space is more than the scale of measurement and the container.

What about the ball? The unpredictable way of the ball, the rhythm of its jumps, tells other stories about the connections of time and space in social life. Whether on the green field of soccer or at the wall of pelota or in the circle of sepak raga, time and space are formed as social dimensions by the rolling and jumping ball. More than that: by the ball, human beings enter into a complex dialogue with each other and with the environment. This dialogue can be a meeting, an encounter ('*Begegnung*', in Martin Buber's (1973) terms) or a disencounter ('*Vergegnung*'), enriching or destructive. Time and space are social dimensions of this dialogue, and this means that they will never be universal but always specific, always cultural. The universal ball-game does not exist and will never exist: ball-games will always be football or tennis, pelota or basketball, sepak takraw or anything else. Maybe the universality of time and space as scientific dimensions is (and always was) an optical illusion? Who are we when kicking the ball? What is our basic spatiality, and what is our basic temporality in movement and social relations? And how are they changing when we change the game, a change that we call history? In other words: show me how you are running, and I can see something of the society in which you are living.

BIBLIOGRAPHY

Bale, J. *Sports Geography* (London, 1989)

Bale, J. *Sport, Space and the City* (London, 1993)

Bette, K.-H. *Körperspuren. Zur Semantik und Paradoxie moderner Körperlichkeit* (Berlin, New York, 1989)

Buber, M. *Ich und Du*, new edn. in Buber, M. *Das dialogische Prinzip* (Heidelberg, 1973)

Dietrich, K. and Heinemann, K. (eds) *Der nicht-sportliche Sport* (Schorndorf, 1989)

Dressen, W. (ed.) *Nilpferd des höllischen Urwalds. Spuren in eine unbekannte Stadt. Situationisten, Gruppe SPUR, Kommune 1* (Berlin, 1991)

Eichberg, H. *Leistung, Spannung, Geschwindigkeit* (Stuttgart, 1978)

Eichberg, H. *Die historische Relativität der Sachen* (Münster, 1987)

Eichberg, H. *Leistungsräume. Sport als Umweltproblem* (Münster, 1988)

Eichberg, H. 'Von Tristram Shandy zu "Marschall Vorwärts". Zur sozialen Zeit der Körper in Sport, Krieg und Fort-Schritt', *Sportwissenschaft*, 1989a

Eichberg, H. 'The labyrinth. The earliest Nordic "sports ground"?', *Scandinavian Journal of Sports Sciences*, 1989b

Eichberg, H. 'Forward race and the laughter of Pygmies. On Olympic sport', in Teich, M. and Porter, R. (eds) *Fin de Siècle and its Legacy* (Cambridge, 1990)

Eichberg, H. 'New spatial configurations of sport? Experiences from Danish alternative planning', *International Review for the Sociology of Sport*, 1993 (Chapter 4 in this book)

Eichberg, H. 'Narrative sociology', special issue of *International Review for the Sociology of Sport*, 1994

Eichberg, H. and Jespersen, E. *De grønne bølger* (Gerlev, 1986)

Eichstedt, A. and Polster, B. *Wie die Wilden. Tänze auf der Höhe ihrer Zeit* (Berlin, 1985)

Foucault, M. *Discipline and Punish. The Birth of the Prison* (London, 1979)

Gaborjáni, P. and Dvorsky, H. (eds) *Hungarian Organic Architecture* (Budapest, 1991)

Habermas, J. *Die Neue Unübersichtlichkeit* (Frankfurt/Main, 1985)

Hall, E. *The Hidden Dimension* (New York, 1976)

Hall, E. *The Dance of Life. The Other Dimension of Time* (New York, 1984)

Hansen, V. *Illustreret Idrætsbog, Vols. 1–2* (Copenhagen, 1893)

Hess, R. *La valse. Révolution du couple en Europe* (Paris, 1989)

Hopf, W. *Soziale Zeit und Körperkultur* (Münster, 1981)

Johnson, M. *The Body in the Mind. The Bodily Basis of Meaning, Imagination, and Reason* (Chicago, Ill., 1987)

Klein, G. *Frauen Körper Tanz. Eine Zivilisationsgeschichte des Tanzes* (Weinheim, Berlin, 1992)

Koselleck, R. 'Fortschritt', in *Geschichtliche Grundbegriffe, Vol. 2* (Stuttgart, 1975)

Lichtenstein, C. and Engler, F. (eds) *Stromlinienform. Streamline. Aérodynamisme. Aerodinamismo* (Zurich, 1993)

Moegling, K. *Alternative Bewegungskultur* (Frankfurt/Main, 1988)

Nitsch, J. (ed.) *Stress* (Bern, 1981)

Oettermann, S. *Läufer und Vorläufer. Zu einer Kulturgeschichte des Laufsports* (Frankfurt/Main, 1984)

O'Neill, J. *Five Bodies. The Human Shape of Modern Society* (Ithaca, NY, London, 1985)

Penz, O. *Bewegungsfelder. Phänomenale Geschichte der Körpertechnik* (Münster, 1987)

Penz, O. 'Sport and Speed', *International Review for the Sociology of Sport*, 1990

Polanyi, M. *The Tacit Dimension* (New York, 1966)

Puig, N. and Ingham, A. (eds) 'Sport and space', special issue of *International Review for the Sociology of Sport*, 1993

Rahbek Christensen, L. *Hver vore veje. Livsformer familietyper og kvindeliv* (Lyngby, 1987)

Sachs, W. *Die Liebe zum Automobil* (Reinbek, 1984)

Schivelbusch, W. *Geschichte der Eisenbahnreise. Zur Industrialisierung von Raum und Zeit im 19. Jahrhundert* (Frankfurt/Main, 1977)

Virilio, P. *Vitesse et politique* (Paris, 1977)

Virilio, P. *Fahren, Fahren, Fahren* (Berlin, 1978)

INDEX

Aalto, Alvar 74

Bale, John 3–21, 23
Benedict, Ruth 34, 35, 37
body culture 4, 5, 7, 8, 11–16, 24, 27, 37,
　41, 42, 71, 72, 74, 76, 90, 92, 93, 97, 98,
　103, 104, 105, 111–15, 118, 120–3,
　126, 128–48, 149, 151–4, 162
body, sanitised 13
body, space 15, 45–67
body–world communion 11
Bogdanovic, Bogdon 73
Brownell, Susan 15, 17, 22–44

Carter, Marshall 23, 24, 28
colonialism 7, 31, 32, 93–7, 100–7, 119,
　128, 132, 137
configuration(al) 4, 15, 29, 34–41, 50, 51,
　54, 60, 61, 64, 68, 69, 76, 80, 90, 92, 95,
　101, 102, 113, 135, 139–43, 146, 147,
　151–4, 156, 158–60, 163
Cosgrove, Denis 7, 8

dance 116–18, 128–30, 136, 137, 141,
　144, 147, 154, 156, 159, 160, 162
disability 10
disciplining bodies 5, 15, 16, 62, 152, 155
Dorn, Michael 10

ecology 15, 28, 31, 32, 34, 41, 70–2, 157,
　158
Elias, Norbert 4, 32, 34, 35, 37–9, 63, 80
elite sport 4, 89, 92, 94, 100, 102, 121, 122

folk-game 3, 103, 132–4, 153; see also
　indigenous games
Foucault, Michel 10, 14, 27, 30, 32, 34,
　36–9, 69, 79

Galtung, J. 7
geometricisation 13
Gregory, Derek 9, 13
Grundtvig, N.V.S 4, 112, 121
GutsMuths, J.C.F 52, 62
Guttmann, Allen 3, 22–4, 28–30, 32–4,
　37
gymnastics 48, 49, 52, 53–7, 60–3, 66,
　68, 70, 71, 73, 75, 76, 78, 91, 95, 104,
　111–13, 116–18, 129, 130, 142, 144,
　147, 152

Habermas, Jürgen 79
Haraway, Donna 9, 14
Harvey, David 7, 13
Hellspong, Mats 116
Hill, Miriam 11
HIV/AIDS 9
Hundertwasser, Friedensreich 73
hybrid 13, 125

indigenous games 12, 13, 48–50, 93, 94,
　101, 103, 104, 128–48

Jahn, Friedrich 52–6
Johnson, Louise 8, 10, 11

Krüger, Arnd 23, 24, 28
Kuhn, Thomas 37

Lacan 13
Latour, Bruno 14
Laws, Glenda 10
Le Corbusier 12, 68, 81
Lefebvre, Henri 13
Lorenzer, Alfred 80

Makovecz, Imre 73, 74

Mandell, Richard 3, 22–4, 28–30, 32–4, 37
Merleau-Ponty, Maurice 11
movement culture *see* body culture
movement house 69, 78, 79, 156

nation(alism) 7, 8, 15–17, 23, 25, 26, 30, 31, 74, 85–99, 101, 103–5, 112, 129–38, 144, 145, 147, 149, 158
New Left 26
New Right/rightist 23–7
Nitschke, August 25, 33, 34, 37, 41, 42

Olympics 7, 12, 16, 28, 30, 59, 89–92, 94, 96–8, 100–7, 129, 132–4, 136, 144, 145, 152, 153, 157, 160

panopticism 14, 73, 79, 151, 156–8, 160, 161
performance 4, 9, 39, 40, 61, 90–2, 122, 123
Philo, Chris 3–21
Pile, Steve 11
Pred, Allan 6

ritual 28–31, 117, 145
Rogers, Ali 7, 8
Rose, Gillian 8, 13
Rousseau, Jacques 61

Seamon, David 11
segmented space 7
Sibley, David 13

situated bodies 11
Soja, Edward 7
Spiess, Adolf 47, 54, 61
Sport Aid 7
sport: achievement 4, 28–31, 34, 40, 41, 72, 80, 81, 94, 96, 102, 104, 122–5, 143–6, 149, 159, 161; architecture 47–83; clothing 56, 57; gender 8–10, 13, 57, 76–7, 102, 105, 118, 120, 138, 141, 144, 145, 156, 158; landscapes 6, 7, 8, 15, 16, 59, 81; modern(ity) 3, 5–7, 13, 15, 28–31, 34, 37, 68, 80, 81, 122, 134, 138, 140, 142–7, 149–63; museumised 17, 70, 136; postmodern(ity) 6, 13, 79, 145–7, 156, 157; premodern 5, 12, 128–48, 151, 153, 154; public 89, 91, 92, 94, 95; representational 7; welfare 17
sports space 15, 47–83
Springwood, Charles 7
stress 153
superstition 65

Teichmann, Frank 24, 26, 30
territoriality(isation) 7, 15
Thrift, Nigel 14
tourism 55, 58, 59
traditional games *see* indigenous games
trialectic/'third position' 4, 16, 22, 25, 28, 80, 88, 97, 98, 120–6

World Cup 7

Printed in the United States
87733LV00003B/343/A